Also by Jeffrey Kluger

Lost Moon: The Perilous Voyage of Apollo 13
(with Jim Lovell; published in paperback as *Apollo 13*)

The Apollo Adventure

Journey

Remarkable Expeditions Past Our Moon and to the Ends of the Solar System

Beyond
Selēnē

Jeffrey
Kluger

Simon & Schuster

SIMON & SCHUSTER
Rockefeller Center
1230 Avenue of the Americas
New York, New York 10020

SIMON & SCHUSTER and colophon are registered trademarks
of Simon & Schuster Inc.

Designed by Deirdre C. Amthor

Manufactured in the United States of America

10 9 8 7 6 5 4 3 2 1

Library of Congress Cataloging-in-Publication Data
Kluger, Jeffrey.
    Journey beyond Selēnē : remarkable expeditions past our moon and to the ends
of the solar system / Jeffrey Kluger.
        p.   cm.
    Includes index.
    1. Space probes—History. 2. Space flight—History. 3. Outer space—Explo-
ration—United States—History. I. Title.
TL795.3.K58   1999
629.43'54—DC21   99-14149
CIP
ISBN 0-684-84765-5

PHOTO CREDITS

Insert pages 1–3, 6–8: Jet Propulsion Laboratory, NASA
Insert pages 4–5: NASA

With love to Alejandra, for keeping me mindful of the magic both at home and on the moons

Contents

Prologue

Linda Morabito was alone in her lab when she discovered that the moon was exploding. Actually, there was some question as to whether it was a moon that was exploding at all. A lot of people—Morabito included—had begun to regard the body more as a fully certified planet, and a lot of *other* people were coming around to that way of thinking, too. But the official position at NASA's Jet Propulsion Laboratory in Pasadena where Morabito worked was that the maybe-planet was indeed a moon, so that's what Morabito generally called it. In any event, it was exploding.

Morabito did not know exactly what she was expected to do if she found out the moon was exploding today. Earlier in the day would have been different, but earlier in the day she wasn't the only one here. Indeed, earlier in the day it seemed that everybody was here: the head of the lab, the head of the department, the head of the whole space agency flown in special from Washington. That was the way it always was on picture days, and there was no reason today would have been any different.

Picture days, it seemed, almost always happened on a Thursday or a Friday, and for a place like JPL, that made sense. When you're showing off the first images of some new world taken by a billion-dollar spacecraft a half billion miles from home, you want to schedule things carefully. Call your press conference too early in the week and the newsmagazines that don't come out until next Monday start to see your stuff as stale. Call it too late and they can't get their stories written up by their Friday-night deadlines. No, if you want your pictures to get anywhere beyond the four walls of the JPL imaging room where the hot pixels sent back from the remote robot probe were first assembled into images, you had to call the reporters in, give them what they needed, and have them on their way no later than lunchtime Friday.

Morabito, of course, was not invited to participate in the press conference earlier today—and she didn't expect to be. Nobody from spacecraft navigation ever was. The way the media saw things, the navigation section's job was merely to get a spaceship (in this case, *Voyager 1*) from some terrestrial Point A (in this case, Cape Canaveral) to some cosmic Point B (in this case, Jupiter) in as little time as possible. Never mind that there were 400 million miles between Jupiter and Florida. Never mind that, at its absolute fastest, the little tin ship that was making the trip would never be able to gun its speed much beyond 35,000 miles per hour, a glacial creep that meant it would need years to get where it was going. Never mind that there were a thousand wrong turns the ship could make in the course of its journey, and if it made just one of them it would spin off into the void, never to contact Earth again. No, the reporters didn't care about any of that. What they cared about were the pictures the ship would be beaming home, and it was the planetologists and geologists from the glamorous imaging team—not the drones from the navigation team—who would present them to the reporters.

This picture day the crowd of reporters gathering to receive those images promised to be a big one, but it was not images of Jupiter itself they were coming to see. The planet, after all, was something of a known quantity. Little more than a mammoth, spherical storm of hydrogen and helium, Jupiter had long since been regarded not so much as a planet at all, but as a sort of failed star. It had the size to be a star and the age to be a star, but it never achieved the critical ignition mass to light its internal fires and actually *become* a star. If the solar system had a blown fuse, Jupiter was it.

What circled Jupiter, however, was another matter entirely.

Buzzing electron-like around the giant world were no fewer than thirteen moons. It was 369 years earlier that the four biggest of those satellites—Io, Europa, Ganymede, and Callisto—were discovered. In later decades, better telescopes added Amalthea and Thebe, Metis, Adrastea, Leda, Lysithea, Elara, Ananke, and Carme. However many other moons there might be was impossible to say, but two years earlier, *Voyager 1* and its sister ship, *Voyager 2*, had been launched toward the outer planets to help find out. What intrigued the scientists and the media following the progress of the ships was not merely the exact number of the moons, but the possibility that something might be going on on them.

The moon astronomers knew best—Earth's moon—was, by even the most generous assessments, a carcass of a world: uniformly gray, uniformly dry, uniformly dead. From as far away as Earth, however, astronomers could see that the Jovian moons were a different matter entirely. There were big moons and small moons, patterned moons and plain moons, brightly colored moons and pasty-pale moons. More important, if remote studies with telescopes and spectroscopes were any indications, there were moons that could have atmospheres, water, and even, perhaps, a spark of internal heat. Put them together, and you had moons that could, in theory, harbor life.

It was the life part that interested the reporters most. And it was the internal heat part that was likely to make the difference. No amount of air, water, and organic molecules was going to be able to do all the clever recombining it had to do to create living organisms if you didn't have something to keep them all warm. Out in the interplanetary provinces where Jupiter lies, however—where the sun looks little bigger than a lit match held across the room and offers little more heat—warmth was not such an easy thing to come by. Any energy sufficient to warm a planet-like body this remote would thus have to come from the interior of the planet-like body itself. A place like Earth is a furnace of a world, with a molten core and viscous mantle generating enough heat to keep volcanoes percolating, geysers spouting, and the very continents themselves floating around on all the geological goo like oyster crackers on soup. If a Jovian moon had even a little of this magma-heated metabolism, it just might be able to cook up something living.

That was the theory anyway, and it was one that the press was eager to learn more about as the first of the *Voyager* ships completed its Jupiter flyby and the late-week press conference got under way. For this initial meeting of the media, the focus would be on what *Voyager 1* had been able to learn about the Jovian moon Io—and even the uninitiated appreciated that this was a good choice. Astronomers already knew that Io appeared to have at least a few wisps of atmosphere and a few riverbed-like gulleys, indicating flowing liquid had once been present on its surface. What's more, the surface of the satellite showed a surprising variety of color—with orange and black highlands broken up by ruddy, rusty plains. A moon with this much apparent chemistry going on was a moon that was capable of anything.

It was with much anticipation, then, that the JPL scientists awaited the first batch of Io photos beamed back by the little *Voyager* ship. When the images finally arrived, however, they were, by most measures, a disappointment. Io might have been a dramatic place—tricked out in all the colors Earthbound astronomers had promised it would be—but it also appeared to be a dead place. Nowhere on the surface of the moon was there visible volcanic activity or any other sign of the underground heating the scientists had hoped to find. Io's trace atmosphere and complex soil might yet harbor enough raw materials to give rise to living organisms, but with the moon's internal fires obviously having flickered out long ago, any native life would have flickered out, too.

Putting the best scientific face on this forbidding world would not be an easy thing to do on picture day, and when the media gathered today it was clear they were disappointed. The scientists stressed—and genuinely believed—that with or without heat, Io was still a chemically fascinating place, that its wealth of organic elements could still teach them a lot about how life evolved on Earth and how it might yet evolve elsewhere. The reporters, however, wanted not chemicals, but critters, and at the same time the scientists were congratulating themselves on making their first successful pass over a world that could keep planetary researchers busy for generations, they found themselves tacitly apologizing to the assembled media for not delivering the organic goods. In four months *Voyager 2* would be making its own Jovian flyby, paying special attention to Io's sister moon Europa. If the last moon had been a disappointment, perhaps this next one—with its brightly reflective surface and its odd, icy rind—would have more to offer.

When today's picture day ended, the chemists and geologists went home for the weekend, and, as they always did at points like this in a mission, temporarily turned control of the spacecraft over to Linda Morabito and the rest of the navigation team. Now that the ship had completed its first major rendezvous, it would be the navigators' responsibility to take its bearings, check its headings, and make sure it was pointed true toward whatever destination the planetary scientists had chosen for it next. The tools the navigators would use for that job would be the same Io images the planetary scientists had just shown off to the press—or *almost* the same Io images.

When a spacecraft tearing along at 7.5 miles per second is taking pictures of a moon that's illuminated by nothing more than a smudge of solar light more than 400 million miles away, exposure time is everything. Leave the shutter of your camera open for too short a period, and the picture you'll get will be nothing but an inky smear. Leave it open for too long, and the exquisitely sensitive light-gathering hardware will gather too much, temporarily blinding itself with the flood of incoming illumination and producing merely a white, washed-out sphere where a picture of a richly textured moon should be. At least half of the Io images *Voyager 1* had beamed down were either overexposed or underexposed this way, and while the blackened, underexposed ones did no one any good, the overexposed ones turned out to be surprisingly valuable.

In addition to gathering in the reflected light of the nearby moon, a spacecraft that left its camera's lens cap off too long would pick up lots of tiny pinpoints of starlight. For a navigation engineer like

Morabito, this was a very good thing, since there was no better way to confirm that an unmanned ship was adhering to its planned trajectory than to check its position against the stars. Each time an overexposed picture came down from the ship, it was therefore passed on to the navigation section, where celestial map readers like Morabito would determine exactly where in the heavens the spacecraft was, compare this with where it should be, and decide if a course-correcting engine burn was necessary.

On the evening of the Friday press gathering, Morabito sat in her JPL lab, studying her computer screen as it flashed its navigation images with their hopelessly fuzzy Ios and their wonderfully sharp stars. Morabito was the last navigator from the day shift still at her desk, but others, working a voluntary night shift, might be punching in soon. After just a few minutes, however, it became evident that they probably needn't bother. From even a cursory glance at the stars in the Io images, it was clear that *Voyager 1* was flying true, with each stellar pinpoint showing up exactly where the navigation charts said it ought to be. Picture after numbingly similar picture told the same encouraging story, when all at once Morabito noticed an image that wasn't so similar. Up near the high horizon of one Io picture, off at the two o'clock position, she spotted a curious bulge in the otherwise smooth disk of the moon. It was like nothing Morabito had ever seen before. The bump was too small to be another Jovian moon peeking out from behind Io, too big to be dust or a pixel glitch in the imaging equipment. It was, undeniably, a part of Io itself.

Morabito dug through a pile of photographic prints for another image taken from the same perspective. The bulge was still there. She found another taken from a few degrees away; again the bulge. Indeed, no matter where *Voyager* was as it photographed Io's facing hemisphere, the curious mound in the moon's surface remained. It was almost as if there was some odd atmospheric aneurysm swelling over the landscape in a very specific spot. But Io had no atmosphere to speak of—certainly not one soupy enough to produce such a horizon-transforming cloud.

Morabito worked with the images throughout the evening, calling picture after picture onto her screen and pulling photo after photo out of the stacks that surrounded her. It was only when the night had passed and the sun had risen that Morabito, bleary and still alone in her lab, realized the full magnitude of what she was seeing. This cloud, she now knew, could be only one thing: a plume from an active volcano—and a huge one. It was rounded like a volcanic plume; it was semi-transparent like a volcanic plume; and it rose over a fixed spot on the surface—exactly like a volcanic plume. What's more, it rose astoundingly high. Judging by Io's diameter, the cloud of underground hellfire had to extend more than 160 miles into space. If the same exhaust blast occurred on Earth, it would roar into the sky thirty times higher than the peak of Mount Everest. While the plume was big, however, it was also wispy, made more of gas than ash. It was so wispy, in fact, that even *Voyager*'s finely tuned cameras were not sensitive enough to spot it—at least when they were operating as they should. When they occasionally misfired, however—such as when they kept their shutters open longer than they ought to—enough light flowed into the imaging system to make the smoke and gas visible.

Morabito nodded to herself incredulously. Io was alive, explosively alive, home to what appeared to be the most titanically huge volcano in the solar system. And only Linda Morabito—with her washed-out photos scattered all around her—knew it.

•　　•　　•

It was not until Monday that the rest of the Jet Propulsion Laboratory learned what Linda Morabito had discovered over the weekend. And it was not for another week or so that most of them believed it. But slowly, the evidence mounted. Imaging experts digitally enhanced the Io pictures, and the horizon bulge only became clearer. Planetary scientists analyzed its chemical spectrum and found it consisted mostly of sulfur and fine particles—just the stuff a volcano would be expected to give off. What's more, other overexposed images from other parts of the moon suggested smoldering volcanoes there as well. At up to nine different spots on the surface, underground heat appeared to be boiling up and blasting into space. Io, it seemed, was a geological pressure cooker, blowing volcanic holes across its own surface like an overinflated beach ball springing spot leaks.

What this meant for life on the moon was impossible to tell. In the vicinity of the volcanoes, surface temperatures appeared to be approaching a shirtsleeves 60 degrees—more than balmy enough for terrestrial organisms to survive. If the right organic chemicals existed close enough to the volcanoes, it was entirely possible that at least some crude forms of life could have emerged.

The JPL astronomers knew it would be at least a year before they would be able to analyze all of the data *Voyager 1* had beamed back and begin to find out for sure. All they could say until then was that the Jovian system was now known to be a hot system—and hot systems were capable of just about anything. In the meantime, *Voyager 2* was still speeding toward its encounter with Europa, and Europa was likely to make things more complicated still.

Earth-based surveys had long since revealed that the Europan landscape was entirely covered by a bright white crust of ice. Moreover, spectral studies had shown that that ice was composed not of methane or caustic sulfur, but of ordinary water. Melt a little of the ice down over even one hot spot on the moon's surface, and you'd have the first ocean known to exist away from Earth. And it was in the oceans that the only confirmed life in the solar system was known to have begun. If Europa had even a fraction of the heat-giving volcanic activity of its sister moon Io, it could be a practical hothouse for extraterrestrial organisms.

As the *Voyager 2* flyby of Europa approached, JPL scientists planned frantically. For this encounter, they concluded they would not be so choosy about the pictures they'd examine. Overexposed images, underexposed images, images that were little more than a Europan shimmer would all be studied. Planetary scientists would begin scrutinizing pictures of the moon when the spacecraft was still months away from Europa, looking for even the slightest suggestion of a volcanic plume.

Realistically, however, they knew that if there *were* volcanoes on Europa, they probably wouldn't reveal themselves so easily. As thick as the moon's ice layer was, even a relatively big eruption would not make it beyond the surface and into the tenuous atmosphere above. Rather, it would stay beneath the frozen crust, heating the lower layers of ice until they turned first into slush and then, perhaps, into flowing water. The only sign the researchers would get that such thermal turmoil was taking place at all would be in the hard ice that still covered the surface. If the Europan crust was pocked with craters, astronomers would know it was an *old* crust—one that had been repeatedly pounded by meteor storms without ever being resurfaced afresh. If the astronomers found a patch of smooth surface ice, however, they would know that in that area, heat was rising from deep underground, turning the ice viscous and allowing its craters to be filled in and troweled over. It was beneath those fresh plains that oceans—and, in theory, life—might exist.

The day of the Europan encounter, the senior JPL astronomers gathered in the main picture room of the Pasadena complex. Torrence Johnson, the head of the imaging team was there, along with chief planetary scientist Larry Soderblom and chief geologist Brad Smith. Also present was Cornell University astronomer Carl Sagan, who had come to witness the encounter with his usual coterie of graduate students in tow. The images this small group would see today would be projected on half a dozen monitors mounted in heavy steel brackets and bolted to the low-hanging ceiling. The scientists crowded into the room early in the morning, waiting for the moment those monitors would flicker to life. Realistically, they knew it might be a long wait.

From *Voyager 2*'s position deep in Jovian space, the Europa images it beamed to Earth would need a full forty minutes to reach the JPL antennas. From there, the signals would be relayed to a mainframe computer deep in a JPL basement, where they would be rebuilt into a picture. This assembly job would take at least another hour—maybe two or three—and only when it was done would the picture be forwarded to the hanging screens in the little room.

At about 8 A.M. Pasadena time, the scientists received word that

Voyager had encountered Europa and had switched on its cameras. Exactly forty minutes later they got word that the data from the first image had arrived in Pasadena. An hour elapsed, then two, and finally, deep into the third hour, the ceiling monitors began to sizzle with static. The men looked up and watched as a circular image began to resolve itself on the screens, slipped a little, then resolved again. A dumbfounded silence fell over the room.

"What the hell is that?" someone finally asked.

"What's the matter with the picture?" someone responded. Instinctively, though, the men in the room knew that nothing was the matter with the picture at all.

On the screens in front of them was the unmistakable image of a bright white world covered by a sort of sugar shell of ice with barely a single crater anywhere on it. The shell was shot through with a tracery of fine fractures and spider cracks, but apart from those hairline breaks, it was practically pristine. There wasn't a planet or moon in the solar system that didn't show at least a little meteor scarring somewhere on its surface. And yet Europa—which could not have been spared the bombardment all of the other worlds had sustained—had erased virtually all traces of it.

There was only one scientifically sensible explanation. Europa, which the astronomers had hoped might be warm in spots, was warm everywhere. All over the moon there must be enough internal heat radiating up to cause the entire ice crust to soften periodically and then re-form itself. And down *below* the crust, things would only be more dramatic. There, the warmth just might be so great the water would never freeze at all. Beneath Europa's frozen rind might be an ocean that girdled the entire globe. There was only one known place in the solar system where such conditions prevailed, and that was in the icy waters beneath Earth's North and South Poles. And those waters, the researchers in the JPL imaging room knew, fairly teem with life.

"The Antarctic," someone in the room murmured to himself. "I'm looking at the Antarctic."

March 1997

Nobody at the Jet Propulsion Laboratory thought it would take eighteen years to return to Europa. Not that there weren't reasons for such a delay, of course. First of all, there were a lot of other missions JPL was considering flying—missions to Venus, to Mars, to Saturn—and all of those deserved a fair share of the lab's time and money. Then, of course, there was that nasty bit of business with the space shuttle *Challenger* back in 1986. When the shuttle exploded, taking seven unlucky astronauts with it, the space community as a whole seized up and shut down. Surviving shuttles sat idling in their hangars; interplanetary probes sat idling in their dust-free clean rooms. It was only in 1988, when the next of the shuttle siblings at last flew successfully into orbit, that the space program as a whole began to bestir itself, too. Given all that, it was only natural that it took a long time to get back to Europa; a full eighteen years, however, still came as something of a surprise. If there was any consolation for the Pasadena scientists, it was that when JPL did return, it was with a dilly of a ship.

The spacecraft that at last took the lab back to the Jovian system was named, fittingly enough, *Galileo,* after the seventeenth-century astronomer who discovered the planet's four large moons. Like its human namesake, the machine was nothing short of extraordinary, a two-part ship consisting of a suicide probe that would plunge into the atmosphere of Jupiter and a far larger ship that would orbit the world, spending at least two years photographing its atmosphere and its swarm of moons. It was not an easy matter traveling the 400 million–odd miles out to the giant world, and *Galileo,* which was launched in the autumn of 1989, suffered its share of breakdowns on the way. In December 1995, however, it arrived at Jupiter, firing off its probe and then settling into planetary orbit. The ship spent its first year in the Jovian system productively, getting its bearings, calibrating its instruments, and sending back a wealth of photos and data as it flew twice by Ganymede, once by Callisto, and once—fleetingly—by Europa. In early 1997 it was ready for its first close pass over Europa, one that would at last allow it to take a good, detailed look at the moon. If any survey of the marbleized world was going to prove the existence of the ocean the *Voyager* astronomers had been rhapsodizing about since 1979, it would be this one.

JPL scientists knew that little would have changed on the four-billion-year-old Europa in the flicker of cosmic time since the lab had last dispatched a ship there, but at JPL itself much was different. Over the course of the last decade or so, the increasing miniaturization of both the space agency's budget and of the equipment it bought had wrought dramatic changes. In the lab's main mission control building, the amphitheater-like room from which the flights of *Voyager* and so many other spacecraft had been run was now dark and shuttered. In its place was a far less grand, far more efficient warren of flight control cubicles in a modest office building on the JPL campus. The mammoth, immobile consoles that had filled the old auditorium had been done away with, too, replaced by prim little desktop computers plugged into the new workstations.

Most important—at least to the researchers on the imaging team—the communal TV rooms where groups of scientists used to gather to whoop and gape as the first grainy pictures from deep space flickered onto black-and-white monitors had been closed up as well. The next time a spacecraft visited a planet or moon, scientists who wanted to view the snapshots it sent home would simply retreat to their individual offices, and there, with the help of their own computers, click a key and call the images up. The pictures, to be sure, would be crisper; their colors, certainly, would be sharper. The scientists, however, would be all alone when they saw them, glimpsing a brand-new world from the same solitary seat at which they answered their mail and made their calls and ate their brown bag lunches. Whether the trade-off was a good one, no one could say.

It was on March 20 when word went around that the latest images from the Europa flyby were on their way. According to the telemetry readings, *Galileo* had barnstormed the moon at a distance of barely 364 miles—closer than any spacecraft had ever come before. The on-board cameras, according to the same data, had successfully snapped hundreds of pictures before the ship flew on, though just how good those pictures would be would not be known until computers on the ground had had a chance to receive and translate the spacecraft's transmissions. Early in the afternoon, it was said, most of this processing work would be done, and by 1 P.M., the hallways and common spaces in the JPL office buildings began to empty. Torrence Johnson, who, as a young engineer, had sat with Sagan and the rest when the Voyager pictures were first fired home, was once again head of the image analysis team. With thoughts of that fine, communal day running through his mind, he watched as the scientists all around him vanished into their offices and then, after a moment, returned to his own.

Sitting down at his computer, Johnson, like dozens of *Galileo* team members at dozens of computers around JPL, did nothing at first but wait, fixing his eyes on the monitor for a sign that the pictures were ready. Almost immediately, a small tone sounded and a tiny icon appeared on the screen. Johnson moved his cursor to the icon and clicked it quickly. In the offices and cubicles all around him, solitary scientists did the same.

On all of the computer screens, a fuzzily luminous landscape began to appear. From the poor definition and the out-of-focus edges, the picture could have been taken almost anywhere—Europa, Callisto, arctic Canada. After an instant the image scanned again and the contours sharpened. No Canada anymore, this was clearly another world. One more scan and that world was clearly Europa.

Johnson ran his eyes over the familiar eggshell landscape of the smooth, hard-frozen globe—a landscape he had first seen nearly a generation ago—and something quickly caught his attention. Near the center of the image, unnoticeable at first, was what appeared to be a small, jagged peak poking up through the surface. It was little more than a shard really, easy to overlook, but the fact that it was there at all was curious. Europa, as nearly as anyone could tell, had no hard topography—at least not any that was visible. Whatever hypothetical features defined its surface lay deep beneath its hypothetical ocean.

Johnson studied the image curiously and understanding quickly flooded in. If he wasn't looking at a mountain and he wasn't looking at a hill, there was only one thing he *could* be looking at: an iceberg. The size was right, the shape was right, the crazy angle at which it appeared to be bobbing was right. Ice, of course, could exist anywhere; ice*bergs,* on the other hand, could exist only in certain places—places that had oceans. Somewhere beneath the hard face of Europa, the frozen mantle must have melted into liquid, and that liquid was apparently churning and surging, cracking the ancient crust above it.

More pictures flashed on the screen now and more icebergs appeared, all glinting the color of fresh snow. There were big bergs and small bergs; upright bergs and tipped bergs. There were icebergs that had shattered into little more than shards and icebergs that had calved away in chunks big enough to fill San Francisco Bay. All of the countless fragments appeared to be locked in place, as if the surface had gone soft just long enough to allow the ice to chip and tip, after which it had frozen up again. But an ice sheet this brittle would certainly crack again, and when it did, the sea would surely spill through it.

Johnson let the full album of images play across his screen, then slowly got up and walked into the hallway, blinking a bit dazedly at what he had just seen. From identical offices up and down the hall, other scientists emerged, wearing identical expressions. Johnson headed straight for the office of Bill O'Neil, the *Galileo* project manager, and caught him before he could even rise from his desk.

"We've got it, Bill," he said simply. "This is the real thing."

O'Neil turned to Johnson and simply smiled. On his screen, Johnson could see, the icebergs were sparkling, too.

•   •   •

Nobody ever called them pilots, and that always galled them a bit. Nobody ever called them explorers, and that rankled, too. When you're building ships in as unlovely a place as Pasadena, trucking them across the country to as unremarkable a place as Florida, then picking them up and flinging them into space—steering them to planets and moons billions of miles away—you'd like to be known as the celestial adventurers you are. Historians, however, never saw things that way.

The problem was that when it came to celestial adventurers, historians had a bellyful. They called them astronauts—and astronauts, as anyone knew, did remarkable things. Unlike the self-styled adventurers at the Jet Propulsion Laboratory in Pasadena, astronauts didn't just build ships, but climbed inside them. Unlike the adventurers at JPL, astronauts didn't fly their spacecraft by remote control, but from the inside of a cockpit, sailing through space by stick, rudder, and the seat of their silvery pants. Unlike the JPL adventurers, astronauts knew that if the ship they were flying turned rabbit on them, spinning out of control and tumbling off into space, they'd never be coming home again. With all that, it was no wonder historians devoted most of their verse to the astronauts alone, remembering the engineers at the little laboratory in Pasadena as mere exploratory afterthoughts.

But the engineers at JPL were more than afterthoughts. While NASA's astronauts never got further than the moon, the robot probes of JPL *started* with the moon. Before human pilots had dipped barely a toe into space, Pasadena engineers were firing off machines designed to crash-land on the lunar surface, soft-land on the lunar surface, serenely orbit the great girth of the ancient moon. When astronauts finally did reach the moon themselves—settling down into the soil with their air tanks and water bags and all the other pulsing, bubbling equipment they needed to keep themselves alive—the lean, fleet ships of the Jet Propulsion Laboratory had already gone elsewhere, heading off into the true trackless depths of the unexplored solar system. What they found there surprised even their inventors.

Earth's own moon—known as Selēnē by the Greeks, Luna in the Romance languages, and simply "the Moon" in uncluttered English—was never really worthy of such a flurry of names. Smaller than Mars, deader than Mercury, it is one more rock in a solar system of larger rocks. The other sixty-two moons circling six of the other eight planets are another matter.

They are strange, almost whimsical places, those other moons, places where the ordinary rules governing heavenly bodies seem not to apply. They're places where volcanoes spew sparkly snow, where rivers run with scalding ammonia, where geysers spout carbonated water, where lakes brim with organic ethane, where fires that burn on one world dust the cliffs of another with ash, where whole globes may shatter into shards and then reassemble themselves. And they are places where life, even now, might be taking hold.

No two of the five dozen or so moons are exactly the same. There are some that are big—bigger than Mercury. There are others that are small—smaller than Manhattan. Some of the moons are red, some are orange, some are black, some are white. One moon is as reflective as a silver platter; another is tarry and dark; a third is both shades at once, with a leading hemisphere black as asphalt and a trailing one white as snow. Like most worlds, the moons are generally round and generally marked by mountains and valleys—but not always. Some are buffed smooth as eggshells; one is as oblong as a great potato. There may not even be exactly sixty-three moons in the solar system at all. There may, in fact, be many more.

Whatever their exact number, the sixty-three or so moons do exist—a second, unknown solar system within the existing known one. For the last thirty-five years, a whole uncelebrated class of explorers has been visiting all of these worlds, sending out probes to run cosmic reconnaissance for the human species back home. The discoveries made by the unmanned ships—as well as by the few manned ones that have followed them as far as Earth's own moon—have been nothing short of remarkable. The stories of how JPL engineers learned to send their robot spacecraft on their extraordinary journeys have been even more so.

Part I

Near

1

A Splendid Suicide

Pasadena, Calif., February 2, 1964

William Pickering had reason to believe that Lyndon Johnson was mad at him. Certainly, the president had never told the chief of the Jet Propulsion Laboratory as much, and it was unlikely he ever would. But what Johnson said and what Johnson thought were often two different things, and there was little doubt that right about now, the president was fed up.

It was only in the last few minutes that Johnson's mood would have turned sour. Up until then both he and Pickering—along with the hordes of reporters following Pickering's work on both of the country's coasts—were in a fine frame of mind. Pickering had not necessarily expected this kind of attention at this kind of hour. Out at his laboratory in Pasadena, after all, it was already after one in the morning. That meant that back in Washington, where the president and much of the eastern press establishment lived, it was after four. Nonetheless, more than a hundred reporters had crowded into NASA's Washington media room to follow a closed-circuit radio account of the goings-on at Pickering's lab, and two hundred more were jostling to get into JPL's own Von Karman Auditorium to listen to the same transmission.

Up until a few minutes ago Pickering was delighted by such unexpected coverage. A few minutes ago, however, things hadn't yet gone all to hell. Now they had, and Pickering would have been just as happy if the reporters would forget all about his little project for

the moment and come back some other time when he had better results to report.

The problem was that most of these people had been coming
back to Pickering for a while now. For those who were still counting,
this was the sixth test JPL had run like this, and it looked as if it was
going to be the sixth one to come to grief. It wasn't just the cost of
Pickering's little adventures that got people's backs up—though the
little adventures didn't come cheap. NASA was paying about $9 mil-
lion for each of the silvery, sixty-six-foot Atlas rockets Pickering periodically needed; another $5 million or so for the twelve-foot,
second-stage Agena rocket that went on top of it; and a good $8 mil-
lion more for the most important piece of hardware of all: the ingenious, ten-foot tall, vaguely cone-shaped *Ranger* moonship that went
on top of both of them. Add the cost of manpower, and Pickering
was burning up more than $25 million every time one of his improb-
able contraptions left the ground.

What really bothered the press, the public, and the president
wasn't just the cost of these Rolls-Royce spacecraft, and it wasn't
just the fact that they were all failing. What really raised hackles was
that they were failing so publicly. All five of the previous *Ranger*
launches had been globally covered events; and all five had been followed by some globally covered disaster, which itself was followed
by a globally covered red-faced press conference in which Pickering
and the rest of the JPL brass tried to stammer through an explanation of what went wrong and pledge—unconvincingly—that they
knew what had to be done to put things right.

For anyone who knew anything about space travel, putting
things right should not have been all that difficult. Pickering's ships
weren't intended to carry people to the moon, after all; that was for
the *Apollo* teams to pull off sometime before the end of the decade.
These ships weren't even intended to make an unmanned but controlled landing on the moon; that was for the engineers building the
still-to-be-unveiled *Surveyor* spacecraft to figure out in the next two
or three years.

No, the *Ranger*s were supposed to beat both of these craft to the
moon to take some preliminary readings, snap some preliminary

photos, and do both the easiest way they could. That meant a crash landing. Launched from Earth on a dumb, cannonball trajectory, a *Ranger* ship would fly all but mutely out to the moon, doing virtually nothing at all for the sixty-six or so hours it needed to get there. Then, in the middle of the third day of its life, when the lunar sphere had grown from a distant aspirin tablet to a genuine arm's-reach world and the speeding spacecraft was roughly fifteen minutes away from annihilating itself on its surface, it would switch on a bank of six cameras and begin taking pictures of the fast-closing landscape, beaming them back to NASA's Goldstone Tracking Station in the Mojave Desert at a rate of better than five images every second. Less than a single second before impact, the *Ranger* would click its shutters for the last time, and then, like a sea captain flinging a bottled message into the ocean the instant before his ship is shot out from under him, toss this final image faithfully home, too. An instant later, it would commit a spectacular suicide on the very patch of lunar soil it had just photographed so faithfully.

Pickering and NASA weren't the first ones to dream up such a kamikaze spacecraft. In 1959—more than three years before JPL's hapless *Rangers* started taking to the skies—Soviet scientists had launched exactly this kind of ship, skeet-shooting the moon with their ingenious *Lunik 2*, a probe that successfully blasted itself to bits in the northern lunar hemisphere and left behind a metal pennant embossed with a coat of arms to commemorate the event. The Soviets had been crowing about their lunar bull's-eye ever since it happened, and in the past two years—after the *Rangers* began flying and failing—Soviet premier Nikita Khrushchev had had himself a ripping good time reminding the world of Russia's successes and clucking in mock sympathy at the struggles of the bumbling Americans.

"The Soviet pennant on the moon has been awaiting an American pennant for a long time," the bumptious Russian liked to say. "It is starting to become lonesome."

Pickering, like the rest of the people at JPL, had gotten sick of hearing about the Soviet pennant, mostly because there probably *was* no pennant—not anymore, anyway. Einstein wasn't kidding when he said that the line between energy and matter is a murky one, and a projectile tearing through space at 6,000 miles per hour would become an energetic object indeed. A mere twenty-pound shot put striking the surface of the moon at such sizzling speed would release a blast of energy equivalent to a suitcase full of TNT. An exponentially bigger, 800-pound spacecraft would be exponentially more powerful, not just exploding, but liquefying—perhaps even vaporizing—the moment it hit the surface, destroying itself and any cargo it may have been carrying. Crash-landing spaceships didn't end their lives in a heroic hail of debris as Khrushchev suggested, but in a sort of ignominious splatter. American scientists tried to explain that to the American public at least a thousand times, but the American public was having none of it. When it came to little *Lunik*, all anybody saw was that glinting, coin-like coat of arms winking down at Earth from Moscow's shiny metal pennant. Heads or tails, the Russians won.

Now, nearly five years after the launch of *Lunik 2*, it looked as if they were going to win again. In the early morning hours of February 2, 1964, *Ranger 6* was more than 240,000 miles from the surface of Earth, just over 1,500 miles from the surface of the moon, and moving at close to its 6,000-mile-per-hour maximum velocity. A speed like that and a distance like that meant that the spacecraft was just seconds away from entering its final, mortal plunge and fifteen or so minutes away from lunar impact. Pickering, like the rest of the engineers at JPL, had spent the last sixty-six hours looking forward to this single quarter hour. *Un*like the other engineers, he would be watching it unfold almost totally alone.

NASA had recently built JPL a spanking new mission control center—or Space Flight Operations Facility, as the space agency liked to call it—a three-story, windowless bunker designed specifically for flights like this. The final preparations of the new SFOF had taken longer than planned, however. While most of the furnishings and electronics had already been installed in the place, the desks and consoles that would fill the mission control room proper had yet to arrive. *Ranger 6*, like all of the earlier *Ranger* flights, would thus be run from JPL's existing, comparatively spartan control center. Pickering himself, however, had decided that in the final minutes of this particular mission it might be wise for him to keep his distance from this older facility, lest his flight controllers, already stooping under the weight of five failed lunar trips, feel even greater pressure as their silent, watching boss looked over their shoulders on the sixth. Instead, he and one trusted lieutenant—Homer Newell, chief of NASA's space sciences division—would retreat alone to the new SFOF's glassed-in viewing gallery, overlooking the eerily empty mission control room, while a public address announcer who was in constant touch with the Goldstone Tracking Station in the Mojave Desert kept them abreast of the progress of the flight.

At just before 1 A.M. Pasadena time, Pickering and Newell took their seats in the new building. At that point, the mission clock at the front of the empty amphitheater read twenty minutes to impact. At the stroke of 15:00, the two men knew, the photography phase of the mission should begin, as *Ranger 6* switched on the power to its video system and began warming up its camera array. Exactly two minutes after that, the six video eyes should actually open and begin transmitting images of the lunar landscape that just 780 seconds later would claim the ship. Alone in the brilliantly lit VIP gallery, Pickering and Newell spent five minutes uneasily glancing around the room, shifting their eyes from the clock to one of the speakers built into the freshly painted and papered walls, and then back again. At exactly the appointed moment, the speaker came obligingly to life.

"Fifteen minutes to impact," the public address announcer said, echoing simultaneously through the old mission control, the new SFOF, and the press rooms in Pasadena and Washington. "Telemetry confirms that the cameras are warming."

Pickering exhaled a deep breath, turned to Newell and gave him a tight nod. Beyond the glass of the VIP booth, on the ghost-ship floor of the empty control room, he could almost see the spectral flight controllers hunched over their invisible consoles. A half minute passed.

"Fourteen-thirty to impact," the public address voice said. "Cameras still warming."

Pickering managed a small smile. The voice coming through the speaker belonged to Walt Downhower, a *Ranger* scientist, and a popular one. Pickering himself had tapped Downhower for this historic, if largely symbolic assignment and he was pleased for both of them that things were thus far going well. Already, *Ranger 6* had lived much longer than any other *Ranger* before it; if it survived another fourteen and a half minutes, it would also have lived much better.

"Cameras still warming," Downhower said again after thirty more seconds. He repeated the call again after another thirty, and once more after yet another. Pickering flicked his eyes at Newell and then at the advancing clock on the forward wall, watching it tick down to the 13:00 mark. Just . . . about . . . now, the half dozen apertures in the half dozen lenses of *Ranger 6*'s imaging system should be flying open. Pickering listened for the call confirming that this had indeed happened, but when he next heard the voice over the public address system, he froze.

"No indication of video," Downhower announced flatly.

Newell turned to Pickering and started to speak, but Pickering raised a silencing finger and held it there. Fifteen seconds later Downhower spoke again.

"Still no indication of video signal," he said. "Cameras are receiving power but Goldstone is receiving no images."

Another fifteen seconds passed. "Still no indication," he said once more.

Pickering lowered his finger and slowly closed his eyes in disgust. If he was getting this news, so too was the press and—somewhere in the White House—so too was the president. At this instant, Pickering knew, the men in the old mission control room just a building or two away would be working furiously, trying to find out just what had blinded the plunging ship and just what could be done to restore its sight in the dozen or so minutes it had left to live. A dozen minutes could be a long time in the life of a spacecraft—even a spacecraft in a 6,000-mile-per-hour death dive—and given the expertise of the men in the room, it was entirely possible they would succeed in getting the problem solved. With a trail of five dead *Ranger*s behind them, however, it was also entirely possible that they wouldn't. Either way, in less than half an hour, Pickering was going to have another press conference to address.

Late 1946

William Pickering always enjoyed the story about the time Wernher von Braun shot a missile at himself. You could say a lot of things about Wernher von Braun—and most people did—but a man who would shoot a missile at himself was a man who deserved at least a little of your respect.

It was just after the end of World War II when Pickering got to know von Braun, at the White Sands missile range in New Mexico. At the time Pickering could not have had less in common with the beefy German. Pickering was born in New Zealand in 1910 and lived there only until 1929, when he moved to the United States to study electrical engineering at Pasadena's California Institute of Technology. New Zealand, unlike most of the other Western and Pacific democracies, had taken its time getting itself electrified, mostly because the dynamos necessary to do the job required oil, and oil required money and New Zealand—which had long gotten by perfectly fine selling the world its butter, wool, lamb, and fruit, thank you very much—wasn't about to bankrupt itself now just to string the home islands with electrical cables. But if New Zealand lacked oil, it did not want for fjords and falls, and once the government was persuaded that it was possible to light the country using nothing more than the power of flowing water, the future of the hydroelectric industry started to look bright.

Pickering, who always had kind of a knack with machines, figured there was a lot of money to be made helping to plug the country in this way, provided he got the right education, and that meant Caltech. The plan was a good one, and it might well have worked out, but barely a month after the visiting student arrived in the United States, the American stock market took a nasty and much-publicized tumble. If the American economy was falling, the economies of much of the rest of the world would be falling further, and in New Zealand, that meant you could say goodbye to electrification. Pickering spent the next six years at Caltech, earning a Ph.D. in electrical engineering and all the while keeping an eye on the job market back home. After graduation, with employment prospects just as bleak in New Zealand as he feared they would be, he decided to settle permanently in southern California.

Even in southern California, it turned out, there weren't that many jobs to be had, but Caltech was willing to help out its own. No sooner had Pickering earned his doctorate than the school offered him an instructor's position, teaching new students essentially the same science he had just been taught. The job didn't include much of the hands-on engineering work Pickering loved, but Caltech sweetened the deal: Whenever the new instructor had time, he was welcome to drive out to the nearby Arroyo Seco riverbed and lend a hand designing some prototype rockets a team of Caltech engineers and graduate students had been fooling around with lately. The place they'd been conducting their research wasn't much—little more than a stretch of dead canyon wash that looked to be more moonscape than landscape. Nonetheless, the instructors and students working there had taken sort of a liking to the little patch of wilderness, and had lately christened it—with decidedly more grandeur than it deserved—the Jet Propulsion Laboratory.

Pickering, who had never much thought about rockets, made a few visits out to the so-called JPL and found that he had something of a flair for the machines being built there. Over the course of the next ten years, he and the other desert designers found they were spending more and more time in their remote riverbed, dedicating themselves to the straightforward task of trying to build new generations of rocket motors that didn't do what most existing rocket motors tended to do, which was blow up. The team had middling success with these efforts, designing both solid- and liquid-fueled missiles that sometimes flew and sometimes didn't, but in neither event went very far. What made this kind of incremental progress hard to take was that even as the Caltech researchers were assembling their firecracker ships, they knew there was one place in the world where the rockets were flying far, indeed: Germany.

During the final years of the Second World War, scientists in both the East and West had watched with a mixture of horror and respect as the German Reich introduced the other nations of the world to its newly invented V-2 rocket. From somewhere in greater Germany—

intelligence had pretty much pinned it down to the northern village of Peenemünde—the military was assaulting hated Britain with a forty-seven-foot-tall, 29,000-pound missile capable of flying over the Baltic Sea to the heart of London and doing a regiment's worth of damage without risking a single German life. The V-2, to be sure, was a brutish machine, a rude and random terror device that may have been good at blowing things up, but did so with little more tactical precision than a hurled stone. To rocket scientists like Pickering, however, such an inelegant device was also a magnificent device. Here was a missile that flew where it was supposed to and *when* it was supposed to, and did both without consuming itself in an unplanned fireball before it got there. If you could somehow put the homicidal purpose of the V-2 hardware out of your head—no small feat, to be sure—the hardware itself inspired nothing but respect. Von Braun, the man who was said to be the rocket's chief designer, inspired almost as much—or at least he started to after the story got around about the business of firing a missile at himself.

During the earliest tests of the V-2, the rumors went, the Peenemünde scientists generally aimed their unproven rockets out over the Baltic. From the safety of their bunkers, they could see the missiles vanish over the horizon and assumed—reasonably—that somewhere in the vast waters off the German coast they were splashing down and sinking. But when the engineers moved their tests inland and aimed instead at firing ranges deep in Poland, they got a nasty surprise: Their rockets went up all right, but inexplicably, they never came down. Somewhere in their trajectory, the fifteen-ton machines were simply ceasing to exist. Von Braun polled his engineers and got all manner of theories to explain the phenomenon, from guidance problems to structural problems to premature detonation of the warhead. What he didn't get, however, was a solid answer. Exasperated, he came up with a way to get one: Scribbling out the coordinates for a particular spot far off in the test range, he handed the sheet of paper to one of his launch technicians and told him to fire the missile at precisely that point in precisely two hours.

Will you be here for the launch? the technician was said to have asked.

No, von Braun reportedly told him.

Will you be back in Berlin?

No, again, came the answer from the chief designer.

Where then? the technician pressed.

Von Braun, the story went, then simply smiled and pointed at the target coordinates he had just scribbled down.

Two hours later the man who invented the V-2 stood at the far end of the ballistic rainbow one of his machines was about to paint in the sky, picked up a field phone, and ordered his team to fire. A few minutes later he saw the black-and-white body of the five-story rocket appear over the horizon, climb high over his head, and start to fall toward him. Von Braun watched and watched as the missile fell and fell, until, shortly before it should have reached the ground, it simply incinerated, breaking into pieces and melting to hot, plummeting slag. Von Braun nodded to himself and walked from the field. The loss of his missile, he now knew, was not due to a problem with its payload or its gyros, but rather to a mere meltdown of its metal skin, which was evidently too fragile to handle the white-hot heat the vehicle generated as it tore through the atmosphere. Von Braun consulted with his engineers, toughened his rocket's hide, and soon had a missile that was fit to fly.

Pickering was intrigued by such a line-of-fire scientist, and in 1946 he learned that he was going to get a chance to observe him up close. Shortly after Germany fell to the Allies, word got around that an arsenal of V-2s had been bundled up like hellish cordwood and shipped to White Sands for research. More remarkably, a detachment of German scientists headed by von Braun had been bundled up in their own way and shipped along with them. Most remarkably of all, Washington was going to ask a few American engineers to travel to the New Mexico desert to study with the Germans, and Pickering was going to be among them.

Over the course of the next decade, with von Braun working exclusively at White Sands, and Pickering—who by now had been named chief of the modest Jet Propulsion Laboratory—traveling back and forth between Pasadena and New Mexico, the Germans and Americans cranked out an impressive line of rockets, some capable of traveling more than 250 miles into the sky. More than 250 miles into the sky, of course, was well into what astronomers thought of as space, and the White Sands engineers liked to think of it that way as well, flattering themselves that they were designing not just rocket ships but starships. The problem with this idea was that while the New Mexico missiles could climb high, they couldn't climb very fast, and without enough propulsive muscle to attain the 17,500 miles per hour necessary to achieve Earth orbit, every putative starship that went up simply fell back down. In 1956, von Braun and Pickering thought they had at last come up with a way to get this problem licked.

Working at his JPL riverbed, Pickering and other Caltech researchers had developed a three-foot tall, solid-fueled rocket that they used principally to study aerodynamics. Out at White Sands, von Braun had developed a slimmed-down and souped-up V-2 that he called a Redstone. In order to study the kind of reentry heating problem that had claimed von Braun's early V-2s, the two scientists had recently combined their inventions, stacking a cluster of eleven of Pickering's little rockets on top of von Braun's big one, and then mounting another three-engine cluster on top of that. The purpose of the improbable assembly was to carry the top stage of the missile as high as it could go, so that when it tipped over and fell back to Earth, it would generate air friction similar to what a vehicle would experience when it returned from space. What the scientists hadn't counted on was just how high and how fast that three-engine top stage would climb. Put one more single-engine stage on top of it, they realized, and it just might be possible to kick a small satellite into orbit.

In as insular a community as the missile community, whispered word of a possible, practical space rocket doesn't stay whispered for long, and in early 1957, General John Medaris, the military commander of the White Sands base, approached Pickering.

"You sure about the design of this rocket of yours?" the general asked.

"Absolutely," Pickering answered.

"It could really get us into orbit?"

"It really could."

"Soon?"

"Very soon."

Medaris nodded slowly. "Impressive," he said. "I wish I could let you fly it."

Pickering looked dumbfounded. "You *wish?*"

"This comes straight from Eisenhower. The Redstone's a military missile, and if we're going to go into space without spooking the rest of the world—particularly the Russians—we're going to have to do it aboard a civilian rocket that can do the same job."

"We don't *have* a civilian rocket that can do the same job," Pickering said.

"Then," Medaris answered, "we'd better invent one."

Whether the scientists and military men of the Soviet Union were actually spooked by the rocketry prowess of the United States was impossible to know. If they were, they didn't let it distract them from their work. On October 4, 1957—just nine months after Pickering's meeting with Medaris—Moscow stunned the West with the announcement that it had just placed a 184-pound robot ball in a 560-mile circular orbit around Earth. The little artificial moon wasn't much—a twenty-inch metallic melon containing instruments no more sophisticated than a thermometer, a radio, and a battery to power both. Nonetheless, the satellite—or the *Sputnik,* as the Russians urged the world to call it—was indeed aloft, speeding around the planet once every ninety minutes and making a conspicuous transcontinental passage over the United States on most of those trips. The rocket used to loft the craft, in the event anyone was concerned, was the huge and fearsome R-7, originally built for the sole purpose of carrying intercontinental ballistic missiles.

In the United States, the scientists answered back fast—after a fashion. Even before Eisenhower dispatched Medaris to talk to Pickering, the president had instructed the Pentagon to order its highly respected Naval Research Laboratory to try to come up with a rocket capable of flying into space. Despite the lab's military pedigree, virtually all of the machines it had ever developed had been used for non-battlefield purposes like weather forecasting and atmospheric study, and Eisenhower was convinced that its space missiles, too, would thus be seen as instruments of peace. In the months Pickering and von Braun's four-stage rocket was sitting idle in its hangar, the Navy scientists had apparently come through, designing a slender reed of a rocket they named Vanguard. The new missile was a disarmingly frail-looking thing next to Pickering and von Braun's burly machine, and next to the Soviets' giant R-7 . . . well, that was a comparison it didn't pay to make.

Appearances notwithstanding, the Vanguards were said to work, and on December 6, 1957, just two months after *Sputnik* was launched, the Navy scientists rolled one of the rockets out to a launch pad at the edge of an old missile range in Cape Canaveral, Florida, uncrated a tiny, shiny satellite of their own, and mounted it lovingly atop the rocket. With news crews covering the event, the engineers backed away, lit the Vanguard's fuse, and watched first in pride and then in horror as the missile smoked a little, steamed a little, rose a few promising feet off the ground and then suddenly ate itself in a bright, white explosion. The spherical satellite, evidently made of sterner stuff than the rocket it relied on, fell to the ground, bounced a few dozen yards away, and came to rest in front of the nearby blockhouse, beeping its idiot signal to the humiliated engineers closed up inside.

The national press howled with laughter at the Canaveral debacle, filling the next morning's newspapers with stories about the Navy and its spectacular new "Kaputnik" satellite. The government only deepened its public relations hole later in the day when a spokesman held a press conference intended to explain away the disaster.

"What do you think was the cause of the explosion?" a reporter asked.

"I'm not sure I'd call it an explosion," the spokesman said.

"What would you call it?" asked the incredulous reporter.

"Rapid burning."

However the government chose to describe its public humbling, it was a humbling just the same, and to Pickering and von Braun, it was an unnecessary one. To the scientists' surprise, General Medaris—the very man who had grounded their far more fit booster in the first place—agreed with them. Several days after the Vanguard disaster, word went around White Sands and JPL that Medaris had been called to the White House to discuss the accident with Eisenhower. When he got there, he wasted little time telling his commander in chief what he thought he needed to know.

If you want a satellite, the general was said to have said, *we can put one up for you.*

What would you use to launch it? Eisenhower reportedly asked.

The Redstone, Medaris answered, betraying not a trace of apology for his military missile. *A Redstone first stage with solid rocket clusters on top of it.*

And you're sure this would work? the president wanted to know.

As sure as we can be.

How long would it take you to launch?

Medaris paused. *Eighty days,* he said. *Give us eighty days and you'll have a satellite.*

At this, the stories went, the president looked at Medaris for a long and stern moment and then nodded. *Eighty days,* he said crisply. Medaris didn't have to be a military man to know that this was an order.

With Eisenhower's okay, Pickering, von Braun, and their rocketeers scrambled to get ready. Of course, just what kind of satellite they'd be firing into orbit if they indeed made the president's deadline was not clear. From the beginning Pickering and von Braun frankly didn't care if they put a bocce ball into space, so long as they put it there successfully. For Washington, however, it was important to maintain at least the appearance that this was more a scientific enterprise than a political one, and the order thus came down that a payload of some kind would be necessary. The engineers complied, cannibalizing a cosmic ray detector that had been built by University of Iowa astronomer James Van Allen for the Vanguard satellite and reconfiguring it to fit on top of their booster. Van Allen, they knew, was a reliable scientist whose instruments would be likely to operate as advertised, and a cosmic ray detector was exactly the kind of otherworldly instrument the politicians would be looking for on this mission. The satellite would be appropriately christened *Explorer 1.*

The booster, shedding its military past if only in name, would be known by the heroic handle Jupiter-C.

The last week in January—weeks before the expiration of Eisenhower's eighty days—Pickering and von Braun, with Van Allen at their side, mated the little *Explorer* to the giant Jupiter and rolled the rocket out to its Cape Canaveral pad. Several days later they announced, probably on the evening of January 31, they would at last attempt to launch it.

When that evening arrived, the beaches along the Canaveral coast were dotted with newsmen, most of whom had been on hand for the headline-making Vanguard pratfall, and none of whom would dare miss this one as a result. Mingling with the reporters were thousands of locals and tourists who had arrived in cars and on foot, and were determined to stay put until the new rocket either flew, or—far more likely—spectacularly failed to fly. Most of the members of Congress, it was said, were following the launch preparations as well, and though few of them had managed to make it here, many had sent press aides who were filing regular dispatches by phone. Even Eisenhower himself, who was on vacation in Augusta, Georgia, and had set the evening aside for dinner and a few hands of bridge with friends, had given explicit instructions that he was to be informed immediately if the Jupiter-C actually left the pad.

To the surprise of nearly everyone on hand, at 10:48 P.M. the Jupiter-C left indeed, bursting to life and rising from the ground in an explosion of hellfire that made its V-2 granddaddy look like a sparkler. Minutes later it vanished into the Florida sky and began speeding toward space, carrying the modest *Explorer 1* with it. Though the crowd that witnessed the rocket's launch was a big one, Pickering, von Braun, and Van Allen were not part of it. After the Jupiter-C took off, it would be impossible to know if *Explorer 1* actually reached orbit for at least ninety minutes—about the time it would take the satellite to complete its first circuit of the globe and at last be in range of the desert tracking station in California's Earthquake Valley. Throughout that time, the project's military sponsors had decided, it would be wise if the three engineers who designed the hardware were not off somewhere on the coast of Florida, but were

within arm's reach at the Pentagon—the better to address a hastily assembled press conference if the mission succeeded, or to tell the De-
fense Department brass what went wrong if it didn't. The men
responsible for creating the rocket that was attracting so much attention tonight were thus nowhere near their machine, but were instead
closed up in a windowless government conference room with General
Medaris and several other officers, following the events by radio.

The three scientists endured the launch in silence—nodding reas-
suringly to the Pentagon men when word arrived that the Jupiter-C
had left the ground successfully—and spent the next eighty-nine or
so minutes smoking cigarettes, glancing at watches, and milling
pointlessly about the room. Finally, at exactly the ninety-minute
mark in the mission—with the stares of the other men in the room
boring into him—Pickering picked up the phone and called his lab.
Frank Goddard, an assistant JPL administrator, took the call.

"Frank?" Pickering said, avoiding the glances of Medaris and the
others. "Anything from the desert?"

"No word yet," Goddard answered.

No word yet, Pickering mouthed to the other men, and tried a
nonchalant smile.

"Are we accurate on time of acquisition?" Pickering asked.

"Should be happening right about now," Goddard answered.

Any minute now, Pickering mouthed.

A silence fell over both the phone line and the room.

"Some launch, huh?" Pickering said into the phone.

"Mm-hmm," Goddard answered.

Mm-hmm, Pickering nodded. He tried the smile again; there
were few in return. Sixty seconds or so passed.

"How late are we, Frank?" Pickering asked.

"Less than two minutes past planned acquisition."

Pickering covered the phone. "Less than two minutes," he said to
the room, and waved a little dismissively. "That's nothing at all.
These things are always rough."

Pickering was telling the truth as far as it went, but he also knew
that while predicting a spacecraft's flight profile was an approximate
science, it could only be *so* approximate. *Explorer 1* might be three

or four minutes late in showing its face, but once it got up to five or six or seven, it became increasingly likely that the satellite was lost. No one said anything for another fifty or so seconds.

"Nothing, Bill," Goddard said at the toll of the third minute.

"Nothing, Bill," he repeated after the fourth.

A fifth minute passed, then a sixth, and then a seventh, each marked by its own grim announcement. Finally, as the eighth minute approached, Pickering closed his eyes. The satellite, he was now certain, had failed. Goddard might not yet be willing to admit it, but Pickering would nevertheless have to announce it, conceding to the generals in the room that, despite his best efforts, the Jupiter-C was just another Vanguard. Suddenly, however, on the other end of the phone line, someone passed a paper to Goddard. Someone else passed him another one, and Pickering suddenly heard a whoop in his ear.

"Acquisition!" Goddard announced. "Acquisition! *Explorer's* talking to Earthquake Valley!"

Pickering pumped a fist in the air. "They've got it," he shouted to the room at large. "We're in orbit!"

In the windowless Pentagon room, pandemonium—or the most that could ever pass for pandemonium in a windowless Pentagon room—broke out. Generals shook hands, slapped backs, and lit cigars while Pickering held one ear, pressed the phone closer to the other, and tried to absorb as much information as he could about the data that was streaming down to Earth from his brand-new moon. Finally, one of the generals caught Pickering's eye, motioned to his watch, and indicated it was time for him to hang up the phone. Pickering did, and almost immediately, he, von Braun, and Van Allen were hustled out of the room, out of the building, and into the back seat of a black government car. With Medaris and two other Pentagon officials pressed in with them, they were driven through the rainy winter night to the nearby National Academy of Sciences.

When they arrived at the academy, Pickering was surprised to see the windows dark and the building all but deserted. Nonetheless, the generals hurried the scientists out of the car, into the building, and down a long, empty hallway until they reached a set of doors marked

"Auditorium." Behind the doors Pickering thought he could hear a hum and murmur. Medaris nodded to the scientists encouragingly and gave the doors a push. Instantly, the hum and murmur turned into a roar. Inside, Pickering could see, were more newsmen than he'd ever imagined could gather in one place at one time. When the reporters caught sight of the party that had just arrived, television lights flicked on like klieg lamps, flashbulbs started to pop like flash paper, and a low, guttural ovation began and built.

Pickering, von Braun, and Van Allen froze in the doorway until Medaris nudged them gently from behind. The newsmen instantly descended on them. Moving in the middle of the crowd toward the front of the room, the scientists realized that they were not quite being borne aloft, but nor were they moving entirely under their own power. Once they reached and mounted the auditorium stage, someone positioned them in a tight group and another fusillade of cameras went off. Someone then changed their pose and the cameras fired once more. Finally, someone handed them a model of the three-foot-long *Explorer* satellite, the satellite that, even now, was circling Earth every hour and a half at speeds approaching five miles per second. The scientists glanced at the model for an instant, uncertain what they should do with it, and then, as if with one mind, they held it aloft. It was this picture that drew the most flashes of all, and it was this picture that would appear in sunrise editions of newspapers all over the world the next day. Pickering surveyed the crowd as the cameras were firing.

Never the same, he thought. *My life's never going to be the same.*

•   •   •

Later that year the government took steps that ensured Pickering would be right. With the country having at last established a toehold in space, Washington moved fast, announcing that it was dissolving its forty-year-old National Advisory Committee for Aeronautics, or NACA, a sleepy federal agency that had been charged with the responsibility of keeping the country abreast of developments in the field of aeronautics. In its place would be the newer, sleeker National Aeronautics and Space Administration, or NASA. Hoping to get itself up and running in a hurry, NASA decided to forgo the tedious business of establishing its own research centers and staffing them with its own scientists. Rather, the agency would simply go cherry-picking among the country's existing missile ranges and rocketry labs, inviting a select few to join the new agency. Those that accepted would continue to be part of the private university or military branch that currently controlled them, but from now on, part of their funds and much of their work would come from NASA.

With the government writing the checks for such an institutional shopping spree, the space agency wasted little time, snatching up facilities from all over the country. Included in NASA's haul were the Air Force's Cape Canaveral missile range in Florida, the Army's ballistic missile test agency in Huntsville, Alabama, and two old NACA centers in Hampton, Virginia, and Greenbelt, Maryland. In June 1959 the space agency also made an offer to Caltech's Jet Propulsion Laboratory in Pasadena, California. JPL was an unprepossessing place—a modest cluster of buildings that had been slapped up in the desert almost haphazardly over the years. But with the right infusion of NASA cash, it might be possible to turn the scruffy little lab into a first-rate spaceport.

Caltech, in consultation with Pickering, accepted the invitation, with but a single condition: While most of the new labs NASA had recruited would be concentrating their efforts on putting astronauts and satellites into near-Earth orbit, JPL wanted to go deeper. The desert lab hoped to build unmanned machines that would not simply circle the home planet, but fly far beyond it, surfing the worlds of the far-flung solar system. If NASA could promise that, the lab would come aboard.

NASA indeed promised, JPL received its membership papers, and before the end of the year Pickering assumed the reins of the newly federalized space center. His engineers' first target, he decided, would be the moon. The ship that would get them there would be called *Ranger*.

August 23, 1961

Even deep inside Cape Canaveral's Hangar AE it was possible to hear the frogs croaking. When you were outside the giant, stamped-tin building that stood in the middle of the old Air Force base, you expected to hear frogs. Indeed, from almost anywhere on the 15,000 acres of mud and sand that was all the cape really was, it was impossible to get away from them. But once you went inside—once you slammed the door of the building where you spent so many of your workdays and tried to concentrate on what you were doing—you expected a little peace. In Hangar AE you rarely got it.

It didn't come as a surprise to most people that the walls of the giant Canaveral Quonset hut weren't quite up to the job of shutting out the frogs. As mission control rooms went, after all, the flimsy building wasn't much. Yes, it had consoles and monitors and headsets like any other control center; some time ago someone had even thought to throw a huge swatch of industrial carpeting down on the raw, concrete floor. But an airplane hangar was still an airplane hangar, and when it came to accommodations NASA didn't appear to be prepared to offer more.

Certainly, the agency was *capable* of offering more. Earlier this year Alan Shepard and Gus Grissom had flown their historic *Mercury* capsules atop their historic Redstone rockets, becoming the first and second Americans ever to travel into space, and during those flights the whole world had gotten a peek inside the glittering mission control room where teams of earnest young engineers nursed the spacecraft off the pad, into space, and safely down again. The men in Hangar AE, however, weren't launching men into space, but robot probes, and for a job like this, a stamped-tin Quonset hut would have to do. Tonight, that job consisted of launching the *Ranger 1* space probe toward the moon.

It had been more than two years since William Pickering and his Jet Propulsion Laboratory had joined the fledgling NASA, and much of that time had been devoted to preparing for this one night. Pickering had spent most of the past twenty-four months shuttling between

JPL and NASA's new Washington headquarters, seeing to both the design of the new ship and the funding that would get it built. His wife, Muriel, had spent most of *her* time learning to tolerate his frequent absences. Muriel was the sister of one of Pickering's old Caltech roommates, and the two had met early on in Pickering's undergraduate career, when she came to campus to visit her brother. Having spent so much time around apprentice engineers, Muriel had grown accustomed to their near-mad absorption in their work, but even this had not prepared her for the pace her husband had been keeping since 1959.

"I'm home a lot more than you think," Pickering used to insist unconvincingly. "On average, I only go to Washington once a month."

"Then most months," Muriel would answer archly, "you beat the average."

Muriel's objections notwithstanding, on this evening, all the coast-to-coast commuting appeared to have paid off. Strictly speaking, the *Ranger 1* moonship Pickering was preparing to launch tonight was not headed straight for the moon, and barring some gigantic miscalculation in trajectory, it would never even come close to it. Where it was headed was much farther into space. From the beginning, the engineers designing the *Ranger* spacecraft realized that the machine they were building had to be a hardy piece of hardware. Leaving the ground aboard the bone-jolting Atlas-Agena rocket, it would be required to survive a rapid acceleration to 25,000 miles per hour and a three-day journey in the killing, −200 degree cold of translunar space, arriving in the lunar vicinity with none of its countless delicate systems either frozen or broken or shaken to pieces. It was an improbable bit of shipbuilding, and in order to prove it could be done, the engineers needed a shakedown cruise, a test flight with a prototype *Ranger* that didn't have to trouble itself with making it to the moon, but merely with getting into deep space, flying about a bit, and making sure its systems worked as they should.

Tonight's flight, then, would be less a finely aimed arrow-shot at the shiny center of the lunar apple a quarter million miles or so away than a crude hammer throw more than 500,000 miles into the void, during which the controllers would send the ship a series of commands and check to see how it responded. After a day or two of these deep-space calisthenics, they would simply abandon the craft to the cosmos.

Pickering, of course, was here for tonight's launch, flown in special from his JPL headquarters out in Pasadena, and from the moment he arrived he could see that his hangar was a busy place. Though the launch of *Ranger 1* was still hours away, the giant Atlas missile was already on its pad, sweating clouds of liquid oxygen vapor. Tucked inside its nose cone like a kangaroo joey was *Ranger 1* itself, transmitting its vital signs back to the control consoles in the hangar, where the nervous scratchings of oscilloscope beams served as a sort of vehicular EKG, reassuring the engineers seated at their monitors that the cargo was healthy.

The signals streaming back from the launch pad were not being monitored in just one control center, but in three. There was the launch blockhouse—located about half a mile from Hangar AE—where Wernher von Braun and a few select members of his Peenemünde team were waiting for the signal to light the engines that would loft *Ranger 1* into the sky. There was Hangar AE itself, where controllers would see to the welfare of the ship from the time the rocket's engine bells cleared the launch tower to the time the spacecraft arrived in space. Finally, there was a third, equally modest control room out in Pasadena, where a remaining group of engineers would oversee the rest of the mission.

Though Pickering supervised the work in all of these facilities, in Hangar AE he was not the most important man present. That person was Jim Burke, the scientist Pickering had tapped to head the *Ranger* program as a whole. Like Pickering, Burke had learned his craft at Caltech, and had first practiced it for the military, building and test-flying battlefield rockets like the Loki, a nasty little needle of a thing that was designed for the job of targeting enemy warplanes and blowing them out of the sky. Burke was an easy, congenial man, and whether or not he enjoyed such a grim line of work was impossible to know. When he showed up at JPL apparently looking for a happier line, however, the lab's bosses quickly hired him on, evidently trusting him enough to put the entire *Ranger* program in his care.

Burke took to his new job readily—all the more so because he'd be performing it among friends. Joining him in the new enterprise would be Bud Schurmeier and Cliff Cummings, two of his Caltech classmates, with whom he'd spent years studying, skiing, and recreationally flying. Cummings would be one of Burke's superiors, overseeing the development not just of the *Ranger* probes, but of any other unmanned moonships JPL might build down the line as well. Schurmeier would be one of Burke's lieutenants, overseeing quality control and development of the *Ranger* spacecraft's on-board systems. Pickering was pleased that three people who would be working so closely with one another were already good friends, but what mattered more to him was that they were also good engineers. Tonight, with *Ranger 1* smoldering on its Cape Canaveral pad, he'd need all their skills.

It was shortly before sunrise when the *Ranger* countdown clock at last reached 0:00 and the Atlas missile carrying the fragile moonship rumbled to life. The controllers inside the hangar could hear a low, almost subauditory roar growing outside and, as if on cue, turned to look over their shoulders. In the control center where the *Mercury* flights were run, huge TV screens at the front of the room afforded a from-the-blockhouse view of the rocket carrying the astronaut from the moment it was fueled to the moment it vanished into the sky. In Hangar AE, where engineers launched robots, not men, the front of the room was nothing but a blank aluminum wall. On the walls at the back of the room, however, high up near the ceiling, was a row of transom-like windows. During daytime launches the fire from even the brightest rocket was not enough to add much light to the bleached Florida sky. But during a night launch the erupting missile more than two miles away caused the entire bowl of the heavens to light up an iridescent orange. A glance at the windows from inside the hangar could thus help reassure the controllers that their rocket was igniting as it should. Now, looking backward, the *Ranger* team saw with satisfaction that the glass was indeed starting to flare the color of fire.

"Liftoff confirmed," Burke said into his headset, turning his eyes

back toward his screen, where the cathode scribble told him by analog inference what the transom had told him directly. "Launch tower
has been cleared."

If the missile machinery was performing as it should, it should
not be long before Cape Canaveral would once again be dark and
quiet. The Atlas rocket would do most of the heavy lifting on this
flight, carrying the moonship all the way up to a near-orbital altitude
of eighty or ninety miles. The spent missile would then fall away, and
the little Agena booster—a relative popgun of a thing housing the
Ranger itself—would fire briefly, climbing to an altitude of 115 miles
and entering Earth orbit. After less than a single circuit of the planet,
the Agena would light again, pushing the *Ranger* up and away from
its parent world and out into deep space. The *Ranger* would then
separate from the used-up Agena and fly on by itself. From the moment the Atlas left the pad, it was clear that the rockets were following this flight profile precisely, and less than fifteen minutes after the
engines were lit, word came back to the hangar that the Agena and
Ranger were in Earth orbit.

Pickering, Burke, and the other controllers allowed themselves a
moment to stand and stretch. From here they could do little but wait.
Their spacecraft had been fired east and would thus have to make
nearly a full lap around the back side of the world before it could
emerge from the west and head out into deep space. The problem
was, on the back side of the world there was no way to stay in touch
with the ship, and it was while *Ranger 1* was in this communications
blackout that its Agena engine would have to ignite the second time
to kick it out on its final trajectory. If the Goldstone Tracking Station
on the front side of the globe reacquired *Ranger 1*'s signal relatively
quickly—say in forty-five minutes or so—the controllers would
know that the Agena had indeed hit the gas and accelerated the
moon probe out into deep space. If it took longer for the ground to
hear the spacecraft's call—perhaps an hour or so—it would mean
that the engine had failed and the craft was marooned in a pokey orbit around Earth.

As with *Explorer 1*, Pickering knew that everything would hinge
on the word that came back from the Mojave Desert. *And* as with

Explorer 1, when that word finally came, the news at first was not good.

"Sixty seconds to planned acquisition," the tracking officer announced to the room at large when the forty-four-minute mark in the mission arrived. "Goldstone reports no contact."

Sixty seconds later he repeated that call, and at every sixty-second interval after that for the next several minutes. With each announcement the mood in the hangar grew darker and darker. Burke slumped in his chair; other controllers alternately paced and sat. Finally, just over an hour after launch—when the tracking officer had made his announcement more than fifteen times—the call that the controllers feared would come, came.

"*Ranger 1* in low-Earth orbit," the Goldstone man said over the phone to the tracking officer. "Agena apparently failed to burn."

"Roger," the officer repeated for the benefit of the rest of the room. "Agena failure. *Ranger* in low-Earth orbit."

Burke cursed silently to himself; the men on either side of him, unconstrained by leadership positions, muttered their imprecations out loud. Pickering, by contrast, felt surprisingly untroubled. The Agena was a moody rocket and everyone knew it. If it failed to perform as it was supposed to for *Ranger 1,* the telemetry that had been streaming down from the ship would tell the engineers what they'd have to fix in time for *Ranger 2.* For now, the *Ranger* scientists undeniably had a spacecraft in space—never mind that it was near-Earth space—and at the moment, it was heading their way. If Goldstone had acquired *Ranger 1* in California, that meant it was currently less than 3,000 miles away and heading east. Crossing the country at 17,500 miles per hour, it should be over Florida in little more than ten minutes. No one yet knew how well the internal systems of the *Ranger* had survived the punishing launch, but it was important to find out. Pickering thought he knew of a way.

On the Canaveral grounds in front of Hangar AE was a rickety, fifteen-foot light tower equipped with a small, hand-operated antenna. Before launch the antenna was pointed toward the pad, allowing it to pick up the signals from *Ranger* that, in turn, produced the vital-sign scribbles on the controllers' screens. After launch, when the spacecraft should be heading moonward, the antenna would be useless since *Ranger* would be aiming its transmissions not toward some hand-cranked receiver standing in the marshes of Florida, but toward the giant dishes of the Goldstone antennas that had been built to receive just such deep-space transmissions. But *Ranger 1* and its fizzled Agena were not in deep space now, and the little antenna that was never supposed to see them again would once again be within range of their signals.

Even as Pickering was contemplating this, Burke seemed to realize it as well. Glancing at his watch, he jumped to his feet and motioned to a nearby controller, and the two of them sprinted outside into the breaking Florida morning. Wading through the swamp grass, they scampered up the metal ladder of the tower and mounted the platform. Burke seized the handle of the antenna—which was still pointed at the now-empty launch pad—and swung the entire assembly back ninety degrees, pointing it roughly skyward. The controller with him then donned a headset, plugged its cord into an outlet on the platform, and squinted into the sky. If Burke's antenna was picking up the spacecraft's signals, the headset would detect them.

"Got it?" Burke asked.

The controller strained to listen, but heard only a soft hiss.

"Nothing," he said.

Burke fiddled with the handle a bit, moving it this way and that.

"Now?" he asked.

"Nope," the controller said, and then instantly reversed himself. "Yes! Got it. It's coming in."

In the headsets, the engineer could hear something that sounded for all the world like a high-pitched Morse code. What it actually was, however, was the sound of *Ranger 1*'s on-board computer beaming down information in the only language the ground-based computers could understand: the mind-numbing strings of ones and zeros that make up a digital code. A larger transmitter aboard the ship would have been able to send the data faster, but *Ranger*'s tiny system could only manage an electromagnetic trickle—so slow a trickle, in fact, that a technician who was familiar with the systems could actually listen for familiar digital patterns and determine if the ship was healthy or not.

"How's it sound?" Burke asked.

The controller closed his eyes and listened. "It sounds good," he said. "It sounds *really* good."

Burke smiled broadly, reached to his right, and opened a night watchman–style box built into the platform. Lifting the receiver, he was instantly in touch with JPL. For the next several minutes the controller with the headset continued to listen to the stream of *Ranger 1* data, reading all he could back to Burke, who, in turn, read it back to Pasadena.

When the spacecraft at last passed out of range and Burke hung up the phone, he hung it up happily. The ship he and the other JPL engineers had built was operating as it was built to operate, and even though this particular moon probe would be staying close to home, the signals indicated that it clearly would have been up to a lunar trip. Instinctively, Burke looked upward as if he could see the invisible probe passing over, though even if *Ranger 1* had still been directly above him, the brightening sky would have made that impossible.

With the break of day, Burke noticed, the frogs had at last stopped croaking.

•   •   •

After 111 egg-shaped orbits of the Earth—covering three million miles and nearly seven days—*Ranger 1* came ignominiously home, tumbling into the atmosphere, bursting into meteoric flame, and sizzling into the ocean. As Burke and Pickering had anticipated, the JPL engineers were able to read through the hieroglyphs of *Ranger* telemetry and dope out the cause of the Agena breakdown pretty easily.

This Agena—and, indeed, all of the Agenas—were birthed by their inventors with a tiny flaw in a tiny valve that controlled the flow of oxygen to the engine's combustion chamber. The valve, the *Ranger* team found, was placed too close to the chamber, causing it to be exposed to a searing blast of heat every time the engine fired. In the case of most Agenas, which were only fired a single time to put a payload into Earth orbit, this didn't present a problem, since all the valve had to do was open once and close once, and the rocket could then be discarded. In the case of the *Rangers*' Agena, which had to fire twice—once to kick the ship into Earth orbit and once to push it out—flash-frying such a delicate component the first time the engine was lit made it likely that there would never *be* a second time. And indeed, when *Ranger 1*'s Agena tried to start itself back up as it orbited the far side of the Earth, the cooked valve was able to flutter open only a bit—allowing just a little oxygen to drizzle into the combustion chamber—before shutting right back down again. A tiny piece of insulating foil would have been all it took to correct the flaw, but *Ranger 1*'s Agena didn't have that foil, so the intended deep-space ship went almost nowhere at all.

Three months later, on November 18, 1961, the same group of controllers gathered in the same hangar to launch essentially the same spacecraft, though this one was called *Ranger 2*. To their dismay, they got essentially the same results. Once again the Atlas performed well; once again the Agena placed the *Ranger* in Earth orbit; and once again the second burn of this second stage went awry. This time the engine lit as it should have—thanks to the tiny piece of protective foil—but the Agena's gyroscopes failed, causing the ship to wobble and tumble like a badly thrown football. By the time the engine shut down, the spacecraft had completely lost its trajectory footing and found itself stuck in a shaky Earth orbit barely ninety-four miles high. Ninety-four miles is practically treetop height by spacecraft standards, and an orbit so low could not be sustained for long. Little more than a day after *Ranger 2* was launched, it came crashing back to Earth.

With two spacecraft kills on its rap sheet, JPL might have benefited from taking a little time to collect itself. It was hardly wise to send a *Ranger 3* off to the dry seas of the moon when *Rangers 1* and 2 had wound up in the oceans of the Earth. Pickering, however, saw things differently. If there was one lesson missile men learned at places like White Sands, it was that the only way to design a rocket that works is to launch lots of them that don't. When one finally goes where it's supposed to go, just remember how you built it and do it that way again. Shortly after the loss of the second *Ranger,* Pickering called Burke, Cummings, Schurmeier, and other ranking members of the lunar team into his office and reminded them of that fact.

"It looks like we're going to have to sneak up on this problem," he told them. "We're not here to design a spacecraft that we know will fly. We're here to design one that we *think* will fly."

"Seems like we ought to be more certain than that," somebody offered.

"When it works," Pickering said, "we'll be certain." The JPL chief smiled at his team and adjourned the meeting shortly after it began. The lab, it was clear, was going to aim for the moon, and the very next *Ranger* was going to be the one to make the trip.

• • •

The engineers at JPL expected to accomplish a lot of things when they designed *Ranger 3,* but building a wooden space probe was not among them. As the liftoff of JPL's next moonship approached, however, that's just what they were preparing to do.

As happy as lunar mapmakers had always been with the *Rangers'* mission profile, lunar geologists had always been less so— and with good reason. A spacecraft beaming back pictures of the moon might do a lot to help you map the lunar surface, but it's absolutely useless when it comes to mapping the *sub*surface. What the geologists had long dreamed of was equipping the ship with at least a crude seismograph that would be able to take at least a few underground soundings of the moon after the ship hit the surface. That seemed impossible, of course, because when a vehicle hits the ground hard enough to vaporize itself, it's hardly likely that any instrument package—let alone one that contains a delicate watchmaker's contraption like a quake detector—is going to live through the impact. Nevertheless, with the announcement that *Ranger 3* was indeed going to head out to the moon, JPL engineers came up with a way to help at least a few of its instruments survive its calamitous landing.

Attached to the top of the new *Ranger 3*—like a sort of giant pom-pom on the stocking cap-shaped body of the spacecraft—was a thirty-six-inch, shock-absorbing sphere. Packed inside the sphere was a sealed instrument package floating inside a sealed envelope of oil. The way the scientists who designed this unlikely device envisioned things, when the spacecraft was just fifteen miles above the moon and falling at 6,000 miles per hour, the sphere would pop free and a small rocket motor attached to its bottom would ignite, slowing its speed to just 80 miles per hour. A good ten minutes after the spacecraft itself had finished snapping its pictures and annihilated itself on the surface, the ball would at last plop to the ground, bouncing several times in the reduced one-sixth gravity, rolling a few hundred feet, and coming to a stop. The floating instrument package inside, which was deliberately designed to be slightly bottom heavy, would then orient itself, switch on its systems, and spend the next thirty to sixty days looking for moonquakes and transmitting what it learned back to Earth.

The contraption was simultaneously ludicrous and brilliant, but it had one small problem: No one had any idea what the ball should be made of. Metals—even the lightest metals—were too heavy and fragile. Plastics and rubbers—even the strongest ones—were too brittle. Someone suggested surrounding the sphere with shock-absorbing air bags that would inflate shortly before impact, but while that might work on a world like, say, Mars, where residual atmosphere could slow the giant balloon down before it hit the ground, on the airless moon even the strongest inflatable material would simply burst on impact.

Finally, Jim Lonborg, one of the lab's more imaginative designers, hit on a possible solution: balsa wood. It was light, it was strong, and most important, it absorbed shock easily—everything the designers were looking for in their seismic ball. Lonborg approached Burke a little warily and began to pitch his idea, but even before he finished, Burke cut him off, laughing.

"Try it, try it," Burke said to his designer. "I've spent half my life building model airplanes. You don't have to sell me on balsa wood."

That day the wooden seismograph sphere went into development at JPL. Not long after, the first working model was mounted atop the *Ranger 3* spacecraft and tagged for flight to the moon.

Just before the spacecraft was shipped from JPL to Cape Canaveral, it underwent one other, far less conspicuous change. At the bottom of every *Ranger* ship were six metal feet designed to help connect the spacecraft to the Agena rocket; attached to each foot were titanium washers that kept the connection snug. Titanium was a good choice for a component like this since the near-indestructibility of the metal would help the tiny rings survive the vibration of launch without coming loose. In fact, so strong were the washers that engineers began to wonder whether they might even survive the crash on the moon itself. In the event they did, it was decided, something else would survive with them.

Inscribed on the washers of this first moonbound craft were the names of Jim Burke and all of the other key engineers who had spent the last three years midwifing its design. The scientists had done the etching themselves—taking turns passing around a vibrating electric pencil in the privacy of their lab as self-consciously as if they had been passing around a bottle of moonshine—and a number of them had taken the opportunity to personalize things even further. Some had inscribed their wives' names; some had included their children's; one had even immortalized his dog. At first, Burke had reservations about allowing the engineers this manifestly unscientific indulgence, but after a while he relented. Planting their nation's colors on the moon might motivate scientists, he realized. But planting their family colors would galvanize them.

•   •   •

Less than five days after it left Pasadena, the air-conditioned van carrying JPL's new, and personalized, *Ranger 3* arrived at Cape Canaveral and backed into a hangar. The Florida launch team would have about six weeks to prepare for what Pickering hoped would be a January 26 launch, and throughout the last weeks of 1961 and the first weeks of 1962, it seemed that they might make that deadline. But barely a week before the time set for liftoff, it began to look as if the new spacecraft would not be going anywhere at all. This time the problem wasn't the *Ranger* itself, and for once it wasn't the cantankerous Agena. This time it was the giant Atlas booster.

The largest organ in the eighty-foot rocket's anatomy was a six-story fuel tank, designed to be filled with explosive kerosene. Resting on top of the tank was a far smaller one holding liquid oxygen that would mix with the fuel and provide it with the flame-feeding atmosphere that would allow it to burn. The stacking of the tanks was a sensible arrangement in the body of so long and narrow a booster, with one small problem. Liquid oxygen is cold—nearly 340 degrees below zero worth of cold—and placing a tank of the frigid stuff on top of a second tank of kerosene was a sure way to chill that liquid, too, turning the already viscous stuff into an unmanageable goo.

To address the problem, the Atlas designers had added a layer of polyurethane insulation to the inside wall of the kerosene tank and covered that with a microthin layer of stainless steel. The design seemed like a good one until barely seven days before liftoff, when pad engineers discovered that in *Ranger 3*'s Atlas, the stainless steel layer had inexplicably sprung a leak and the insulation inside was sopping up kerosene like a giant sponge. The problem was clearly the fault of North American Aviation, the booster's San Diego–based designer and builder; but now North American's problem had wound up in the collective laps of the JPL engineers.

The moment Burke heard about the leak, he phoned Cape Canaveral, alerted the launch team that he would be flying in from Pasadena, and asked to meet with three people: Kurt Debus, the Peenemünde veteran who ran the Florida launch complex; Major Jack Albert, an Air Force officer whose job it was to certify all Atlases either fit or unfit to fly; and John Tribe, a gifted young booster engineer who had nowhere near the seniority of the other three men, but made up for it in raw missile smarts. The group gathered in Debus's office, and once they were there Burke got right to business.

"So what do we do with this leaky missile?" he asked straightaway.

"Not much we can do here," Albert said. "This is North American's job."

"Which means we pull it back from the pad—" Burke began.

"—and ship it back to San Diego," Debus finished.

"And how long for the fix?" Burke pressed.

Debus turned his palms up. "Two months. Maybe three."

Albert and Burke whistled softly. With two *Ranger*s already lost and NASA starting to ask when it could expect some results from the millions of dollars it was spending for each of these temperamental ships, two months might as well have been two years. Burke started to speak, but Tribe interrupted him.

"Don't ship that rocket anywhere," he said.

"North American isn't going to come here and fix it for us," Albert objected.

"They don't have to," Tribe said with an emphatic nod. "*I'll* fix it here."

The three senior men stared at the junior engineer for a moment. "John," Burke said, "you can't do that without taking the tank out of the rocket, and it takes the designer for that."

"We don't need the designer," Tribe said, "if I go inside the tank."

The plan Tribe proposed was a radical one. At the bottom of the Atlas's kerosene tank was an eighteen-inch hole where a bolt ring held one of the engines in place. If the tank was drained and the bolt removed, an engineer—at least a wiry engineer like Tribe—could wriggle inside. The atmosphere he found there would be poisonous, since no amount of draining could remove the toxic vapors six stories of kerosene would leave behind. But with the help of a gas mask, a miner's flashlight helmet, and a carpenter or two, he could build a wooden, ship-in-a-bottle scaffold, climb the sixty feet to the top of the tank, and perform whatever repair work needed to be done. When it was complete, the scaffold could be disassembled and extracted and the tank refilled.

"How long would this fix of yours take?" Burke asked.

"A week or so," Tribe said, and then added pointedly, "seven weeks less than the contractors."

Even if Pickering and Burke could argue with Tribe's plan, they couldn't argue with his math. Within days *Ranger 3*'s Atlas was rolled into a hangar, its engines were detached, and John Tribe and his pair of helpful carpenters were crawling around in its innards. One week later, as promised, the tank was once again fit to be fueled, and the surgically repaired Atlas was rolled out to the pad. On January 26 it took off for the moon. Like the *Ranger*s that preceded it, it would never get where it was going.

•   •   •

When flight planners plot a path to the moon they are always conscious of the fact that they're aiming at a moving target—and a fast-moving one. The moon needs less than twenty-eight days to inscribe its 1.5-million-mile circle around Earth, which means that it covers more than 53,000 miles every day, or more than half a mile every second. For that reason, terrestrial rocketeers can't simply point a spacecraft to the spot in the sky where the moon is today, but rather, where they calculate it will be in about three days. If the spacecraft takes it into its head to try to speed things up, all it winds up doing is arriving at the designated rendezvous point before the moon itself shows up, and then whizzing past into trackless space. If it moves too slowly, the moon will get there first and be long gone by the time the slowpoke ship arrives. For this reason, launch directors trying to fire a spacecraft on a lunar trajectory had to make it a point to keep an especially tight rein on their speeding rockets.

On January 26, 1962, the day *Ranger 3* was launched, it was clear the Atlas was fighting the bit. Just minutes after liftoff, the incoming telemetry indicated that the engines were burning far too hot, adding unwanted speed to its climb. Burke told his controllers to radio up commands instructing the booster to throttle back a bit, but the booster ignored them and raced on. When the out-of-breath spacecraft arrived in orbit, it was oblivious to the excess velocity it had attained, and before the controllers could countermand its programming, it lit up its Agena and headed out to the moon at the same too-high speed.

Watching from his console in the center of the room, Burke was stunned at what he was seeing. The first and most obvious thought was that the problem had something to do with the improvised fix John Tribe had dreamed up for the Atlas, but Burke knew this couldn't be. Tribe's fix was a simple thermal fix; this problem was caused by something far more complicated. Just what that thing was, though, was a puzzle to be solved later. More pressing now was determining just how much damage it had done. If *Ranger 3* was only going fast enough to miss the moon by a few thousand miles, it might be possible to fire the probe's own pint-sized engine at some point during the three-day trip, slow things down a bit, and hit the lunar target after all. If the ship was going to miss the moon by 10,000 or so miles, however, no amount of corrective action would be much help. It would take the JPL tracking officers several hours to make these calculations precisely, and since the Hangar AE team was scheduled to hand control of the spacecraft over to the Pasadena team at this point in the mission anyway, Burke promptly hopped a military plane for the West Coast.

When the *Ranger* project manager at last arrived at JPL late that day, he headed not for the control center, but for a nearby support room where Marshall Johnson worked. Johnson was JPL's trajectory shaman, and all day today he was holed up with his cabinet-sized IBM computer, feeding it punch cards and waiting for it to spit back data on the flight path of the racing *Ranger*. As soon as Burke arrived in Johnson's work area and set eyes on the scientist, he could see that the news was not good.

"Where do we stand?" Burke asked.

"Spacecraft is high and fast," Johnson said.

This did not help Burke; he knew *that* already. "And?" he asked.

"And we're going to miss the moon by twenty thousand miles."

Burke quietly cursed—and then did what every missile man does when he finds that his vehicle is not going to be able to perform its primary mission: He began thinking of a secondary mission. As it turned out, Pickering—who had gotten Johnson's report even before Burke—was already coming up with one. While *Ranger 3*'s trajectory ruled out a lunar impact, it did not rule out remote photography. In two and a half days the spacecraft would move closer to the moon than any American spacecraft ever had before, and with a carefully planned firing of its engine, that gap could be closed even further—if never completely. All the *Ranger* team would have to do was nudge the ship as close to its lunar target as they possibly could, time its cameras to fire just as it was flying by, and they could still deliver an album of lunar photos the likes of which American scientists had never before seen.

Burke, Pickering, Johnson, and the trajectory engineers sat down in the *Ranger* control room to plan the engine maneuver and discovered, to their delight, that it would be a fairly uncomplicated one. If they simply ordered *Ranger 3* to roll a bit so that its flat base faced forward, and then fired its main engine for a few seconds, they could change its speed by eighty miles per hour. Eighty miles per hour might not have been much for a ship that was currently traveling about three hundred times that snail's pace speed, but factored out over the two-plus days of flight time, this was enough to bring it at least 6,000 miles closer to the moon and improve its picture-taking resolution dramatically.

Late that day Burke's assistants ran the numbers over to a communications officer, who radioed them out to an engineer at Goldstone, who, in turn, radioed them up to *Ranger 3*. A few minutes later word came back that the wayward spacecraft had done what it was told, the engine had fired, and the ship, even now, was easing itself moonward.

At JPL, the relief was palpable. Not only had the *Ranger* team salvaged at least part of their spacecraft's picture-taking mission, they had also planned and executed one of the most delicate bits of deep-space flying ever attempted. The flight team exchanged congratulations all around, but the self-kudos turned out to be short-lived.

The morning after the engine burn Burke got up early and went straight to the control center to meet with his flight dynamics officers and check on the overnight trajectory of his ship. When the men looked at their data, they were stunned. According to the numbers coming down from the spacecraft, *Ranger 3* wasn't moving closer to the moon, but farther away from it. More disturbing was the eerie *way* it was moving away. The ship wasn't wandering randomly off course the way it would if a gyro had blown or a thruster had inadvertently fired. Rather, it was moving at the precise angle and speed it had been instructed to, just 180 degrees in the opposite direction. It was as if JPL had told the ship to make a very precise right, and it had instead made an equally careful left.

Burke suspected he knew just where the problem lay. Collaring two of the trajectory analysts, he tore over to the spacecraft assembly building and began ripping through the programming books that contained the spacecraft's navigational codes. In short order he found what he feared: A single digit out of all the flyspeck numbers that flecked the pages had been inverted, causing a +3 to be entered into the spacecraft's brain where a −3 should have been. No matter where the ship was told to fly, it would thus fly exactly the other way. Burke dropped his head into his hands.

"Talk to Goldstone," he said wearily to one of the assistants. "Tell them to tell the spacecraft to change this"—he pointed to the +3—"to this"—he pointed to the −3. "And tell them," he added, "to do it right now."

The assistant obeyed, but even *right now,* Burke suspected, was probably much too late. The next day *Ranger 3* sped past the moon at a distance of 23,000 miles and tumbled into orbit around the sun, becoming, in essence, a tiny, 800-pound planet. Shortly afterwards, it vanished from JPL's tracking screens altogether without ever having returned so much as a single picture.

•   •   •

The Soviet Communist Party meetings in Vladivostok were usually not very lively affairs. The apparatchiks would troop into the dreary meeting hall for two or three days of speeches, listen as this or that new five-year plan was unveiled, and then troop home again, little more enlightened than when they came. At some point, Premier Khrushchev usually made an appearance, and this generally picked things up a bit, but even Khrushchev had trouble giving what was essentially a prefab pep rally any appreciable pep. On January 30, 1962, however, the premier was in a frisky mood. Another American spaceship—this one they had called *Ranger 3*—had just gotten lost on its way to the moon, and this time it appeared to have gone so awry the American scientists were unlikely ever to hear from it again. When Khrushchev took the stage for his scheduled address, he couldn't resist speculating about the reason.

"Why does the United States try to launch a rocket to the moon but fail to hit the target?" he asked with playful bewilderment. A few members of the audience laughed self-consciously.

"It's not that they haven't got the dollars," he said, "in fact, they have lots of them." A few more audience members laughed.

"It's not that they haven't got any scientists," he went on, "for they have those, too. Then what are they lacking?" he asked. Before anyone in the audience could shout out an answer, he jumped in himself.

"They lack the regime which exists in our country," he sang out triumphantly. "Socialism," the premier declared, "is the only reliable launching pad from which to launch spaceships. Here there is no anarchy of production, no fierce competition, no class antagonism which hinder the necessary concentration of effort necessary to solve scientific problems."

It was familiar stuff to the people in the hall, but it was exuberantly delivered. Khrushchev, in fine form, went on this way for a while, whipping the otherwise somnolent crowd into what passed, at a function like this, for a frenzy. The next day, the complete text of the hugely successful speech was released through TASS, the government-controlled news agency. The day after that, Washington circulated it to NASA; the following day, NASA passed it on to JPL. In Pasadena, most of the *Ranger* officials made a point not to read a word of it.

•　　•　　•

For the JPL scientists, with three dead *Ranger*s on their hands and a tab for the failed moon flights that, including manpower and R&D, was already approaching $100 million, the answer to bad P.R. was to launch again, and to do it soon. So far, nobody in Washington was talking about pulling the plug on the entire troubled program, but that was certain to change. The longer the *Ranger* project went without any tangible results, the more fixed its reputation as a waste of limited space funds would become. The next time NASA's budget came up for review, it would be those deadweight projects that would be the first to go.

It didn't take long to pinpoint the flaw in the guidance program that had led to the excessive speed in *Ranger 3*'s Atlas, and it took even less time to fix it. As soon as it *was* fixed, *Ranger 4*, freshly trucked in from Pasadena, was hoisted atop the new, modified rocket. Just two months after that, on April 23, 1962, it blasted off its Canaveral pad and headed toward the moon.

At first, the flight of *Ranger 4* looked not just promising but downright flawless. The repaired Atlas flew straight and true; the temperamental Agena fired perfectly to put the moonship in Earth orbit; an hour later it refired on cue to kick it out into deeper space. At their consoles in Hangar AE and JPL, the controllers looked at one another with wary, almost incredulous smiles. This was exactly—*exactly*—the way they had built this cursed ship to operate. Indeed, so well was it working today that a short while later, when Marshall Johnson's preliminary trajectory data came down, the flight dynamics officers realized that their jobs were almost done. *Ranger 4*'s aim was so good that without so much as a tweak from its thrusters or a breath from its main engine, it would strike the moon on precisely the day it was supposed to, at precisely the right time and precisely the spot. Even as this cheering news was being passed around JPL, however, the spacecraft was preparing to breathe its last.

All of the *Ranger* moonships were electrically connected to the rocket that carried them by a simple arrangement of plugs and bolts. Ringing the top of the Agena were half a dozen ordinary-looking electrical sockets, and protruding from the bottom of the *Ranger*s were half a dozen pairs of corresponding prongs. When the moonship was mounted on top of its Atlas-Agena, it was thus literally plugged into the rocket, and then secured in place by a series of explosive bolts. After launch, when the *Ranger* had been fired out of

Earth orbit and the Agena's work was done, the bolts would deto-
nate, the ships would separate, the spent Agena would tumble away,
and the spacecraft's on-board batteries and solar panels would take
over the job of providing it with power.

That was the way the system was supposed to work anyway, and
on *Ranger 4* that was the way it *started* to work. But when the spacecraft and Agena separated, a few flecks of foil were set adrift between them. By pure, free-floating randomness, one of the flecks
happened to settle atop two of the electrical prongs protruding from
the bottom of the *Ranger,* and as soon as it did, it completed a killing
circuit that instantly tore through the entire body of the ship. Like an
electrocuted man, *Ranger 4* shook, cooked, and slowly began to sizzle and die.

On the ground, the calamity played out quickly. One moment a
healthy stream of telemetry was pouring down to the happy controllers
in the Earth-based tracking stations; the next moment there was nothing at all but a tiny signal from the tiny transponder buttoned up inside
the balsa wood ball. Since the battery that was intended to power the
seismograph on the surface of the moon was separate from the ship's
power source, it was unaffected by the system-wide short circuit. As
the suddenly cold, suddenly dead *Ranger 4* sped away from Earth at
25,000 miles per hour, this little, muffled voice was all that was left to
mark its transit.

Pickering was in the JPL control center when the spacecraft
telemetry suddenly vanished, and for the better part of the day he,
Burke, and the rest of the controllers concocted different combina-
tions of commands that they hoped might awaken the sleeping ship.
Throughout that time JPL issued only the vaguest mission status
statements to the reporters crowded into Von Karman Auditorium,
but at 6 P.M. or so the media people would expect—and be entitled
to—some kind of end-of-the-day wrap-up. It fell to Pickering to tell
them that the ship they were following so closely had flatlined. Walking into Von Karman, he pushed his way without comment through
the reporters who had been gathered there for most of the day,
mounted the stage, and drew a weary breath.

"It looks like we have a fairly significant problem with the spacecraft," Pickering said a little blearily. "We don't really know how major the problem is."

"What does the telemetry *say* the problem is?" a newsman asked.

"We're not receiving any telemetry," Pickering confessed.

"Is the spacecraft dead?" another asked.

"Not completely."

"Has its guidance system been able to align itself with the sun?"

"No."

"Have its computers been able to communicate with Earth?"

"No."

"What does that mean?"

Pickering paused. "That," he said with a sigh, "indicates that the trouble is deep within the brain of the system."

Pickering was being as honest with the reporters as he could be, and over the next day, as his unconscious ship showed no signs of awakening, he became increasingly candid, conceding to the media that this *Ranger,* too, was likely to be a failure. A dead pod heading for a pointless lunar impact was not the stuff of headlines, and as *Ranger 4* proceeded all but mutely through its two-and-a-half-day journey, Pickering anticipated that the media's interest in it would vanish all but entirely. Unexpectedly, however, the flight of the disabled ship became perversely fascinating to the press. The heartbreakingly hopeful, chick-like peeping of the balsa wood ball's 50-milliwatt transmitter continued to register on Goldstone receivers even as the ship receded farther and farther into the void. As it did, both the reporters and the public, suckers for hard-luck cases, demanded more and more information about its progress. Nobody pretended that the mission would return any good science, but as long as the trajectory held steady, the fact remained that before the end of the week, an American spacecraft—even if it was an American spacecraft sunk deep in an electronic coma—was going to strike the surface of the moon, and that was an event worth noting.

Pickering and the JPL public relations team did what they could to satisfy this surprising national interest, tracking the steady chirps of the seismograph ball and issuing regular updates on the spacecraft's status. Trajectory data gathered by the Goldstone station in the Mojave indicated that the ship would strike the moon shortly between 4:30 and 5:00 A.M. on April 26. A slight list that had appeared in its flight path would now cause it not to hit the lunar sphere head-on, but to soar just past its western edge. This approach would nevertheless be so close that the ship would still be captured by the moon's gravity, arc around its far side, and strike the surface just out of view of Earth. If the peeping of the transponder stopped at 4:47 A.M. when the ship vanished over the western horizon and did not resume some forty minutes later—the amount of time it would take it to reemerge over the eastern side if it did not crash—the JPL men would know they had hit their lunar target. Two and a half days after *Ranger 4*'s launch, Pickering, Burke, Cummings, Schurmeier, and several other JPL officials flew to Goldstone to follow their spacecraft's descent.

When the JPL team arrived in the desert, the sun was still not up and the Goldstone facility was lit by little more than the blinking red lights at the base of its white, wide-mouth antennas. In an auditorium inside, technicians had set up a large television monitor filled with a live image of the moon, captured by a sixty-inch telephoto lens aligned with one of the receiving dishes outside. *Ranger 4*—a distant, ten-foot dust mote—could not possibly show up on the screen, but a superimposed crosshair was able to pinpoint its location as it fell closer and closer to the edge of the moon. A speaker placed next to the indoor monitor broadcast the peeping of the transponder as it called tenaciously back through the growing, deep-space static. A similar system had been set up in a nearby press room for the small crowd of newsmen who had also managed to find their way to this normally forgotten outpost at this normally forgotten hour.

As the JPL scientists settled into their seats in the Goldstone auditorium, *Ranger 4* was just a few thousand miles from the moon. As it drew closer and closer, the gravity lasso that had snared it would pull harder and harder and the ship would begin to move faster and faster. The peeping of the transmitter would continue at the same speed and frequency no matter how fast the *Ranger* was going, but the shape of the electromagnetic signal that was carrying it would change dramatically. Since radio waves tend to stretch and flatten as the transmitter emitting them moves away from the antenna receiving them, all the Goldstone engineers had to do was measure this change and they could measure the spacecraft's velocity, too. As the peep-peep-peep sounded hypnotically through the room and the crosshair on the TV closed in on the lunar limb, an engineer approached with a readout of the transmitter signal and handed the single sheet of paper to Cummings. The transmission, it was clear, was indeed flattening out and the spacecraft, by implication, was indeed in its death plunge.

"Oh, baby," Cummings muttered as he scanned the numbers.

At the front of the room, a large mission clock indicated that loss of spacecraft signal was now just two minutes away. Two minutes melted to one and one melted to just thirty seconds. Nearly 240,000 miles away, Ranger 4 was climbing to its maximum speed and falling to within a few miles of the surface. Then, just fifty seconds after the clock in the front of the Goldstone auditorium flashed 4:47 A.M.—almost the exact moment the trajectory experts had predicted—the peeping from the spacecraft vanished as suddenly as if it had been cut off by a knife and the room was filled with the soft hiss of translunar space. Pickering glanced briefly at his watch and nodded a silent farewell at the screen. Even without waiting the formal forty minutes, he knew he'd never hear from his Ranger again.

•   •   •

Almost from the moment the dead Ranger 4 struck the lunar surface, the JPL press apparatus began cranking itself up to start recasting the failed mission as at least a qualified success. To the delight of the lab, the press seemed more than willing to agree with this happy interpretation of things. Paper after paper hailed the accuracy of the rocket and the precision of the flight, and overlooked, at least for the time being, the death of the payload.

But if the American news community was willing to turn a deaf ear to failure, Khrushchev wasn't. Shortly after Ranger 4 died, the

Soviets were enjoying their National Railway Day celebrations, when the party leader took another opportunity to tweak the West. Delivering a speech to a group of transport workers, he suggested that maybe, just maybe, the Americans' latest spacecraft hadn't hit the moon. After all, he said with a conspiratorial twinkle, three straight *Ranger*s had already failed, and with a fourth one now added to the toll, perhaps the scientists on the other side of the ocean just cooked up a story of a lunar impact to save global face.

This time Pickering was riled. Attack his *Ranger*s if you want; attack his Atlas; you could even attack his engineers. But don't question his truthfulness. Pickering was a missile man first, and when a missile man tells you he hit his target the way he was supposed to, you can be absolutely certain he did. The day after Khrushchev's speech, a curt Pickering delivered an address of his own. Calling a morning press conference at JPL, he waited until the reporters were assembled in the media briefing room and then strode in silently. Mounting the stage, he began his address without a word of preamble.

"On April 26, 1962, at 4:47 and 50 seconds A.M.," he said, "*Ranger 4* was tracked by the Goldstone receiver as it passed the leading edge of the moon. At 4:49 and 53 seconds it crashed on the moon at a lunar longitude of 229.5 degrees east and a lunar latitude of 15.5 degrees south."

Pickering stepped away from the microphone and looked out at the reporters in front of him for a long, resolute second. No one, it was clear, was likely to ask a follow-up question.

•　　•　　•

In the weeks following the death of *Ranger 4,* Jim Burke and William Pickering went in dramatically different directions: Burke decided to go to Europe; Pickering decided to go to Venus. By any measure, Pickering's decision was the safer one.

Though the public and the press did not always appreciate it, JPL was in the business of doing more than just flying to the moon. Nearly eighteen months earlier, in February 1961, the Russians had launched the *Venera 1* spacecraft on a 224-million-mile journey to

Venus. The spacecraft was not much—a tin can of a thing equipped with some primitive cameras and deep-space sensing equipment. The trajectory it flew was cruder still: Fired out of the clean, orbital circle Earth inscribes around the sun, it spiraled slowly in toward Venus, but was never intended to land on it. Rather, it crossed Venus's orbit and flew fleetingly past the cloud-covered world at a whopping distance of 62,000 miles. The pictures the ship could snap were thus little better than what could be captured from the window of a speeding car—and a remote speeding car at that. Nevertheless, snap pictures it did, and a few days later these fuzzy portraits were proudly flashed around the world.

On its face, the feat ought to have seemed impressive, and to people unacquainted with the business of space travel, it indeed was—certainly more impressive than flying to some close-to-home target like the moon. Scientists who made it their business to undertake such flights, however, saw things differently. If human beings were ever going to follow robot probes into space, flying to particular, predesignated spots on the surface of new worlds, they were going to have to master a brand of pinpoint piloting never before even contemplated. To learn this kind of flying, you're a lot better off rehearsing with the close-up, bull's-eye missions the *Ranger* ships were flying than the long-distance miss *Venera* achieved. While American scientists understood this, however, the American public often didn't seem to. And it was the American public that ultimately paid the bills, applied the pressure, and decided if NASA was going to exist at all. If the Soviet Union was flying to Venus, Americans would insist on going, too—and it would be William Pickering's JPL that would be expected to take them there.

As distractions went, the prospect of attempting a mission to Venus was a surprisingly welcome one. After the spectacular failure of *Ranger 4,* JPL was in no hurry to launch a *Ranger 5*. The lab's scientists sniffed out the cause of the *Ranger 4* flop pretty easily and came up with an attractively straightforward solution. On future missions the electrical connection between the spacecraft and the booster would be redesigned, with the plugs repositioned to protrude from the top of the rocket and the sockets now built into the bottom of the spacecraft. Any stray piece of foil that contacted the plug prongs would thus only short out the empty Agena, which by that point in the mission would have been jettisoned as junk anyway.

Uncomplicated as this reengineering work seemed, conducting it would nevertheless take six months or so, meaning that no *Ranger* could set out for the moon until the late fall of 1962. And that left a brief window for Venus. In order to get this first interplanetary mission off the ground as fast as possible, JPL engineers decided not to develop a whole new type of spacecraft to make the trip. Instead, the Venus ship would be little more than a modified *Ranger* ship with longer-range cameras, deeper-space sensing instruments, and no seismograph at all. The name, too, would be changed to a far-more vagabond *Mariner.*

In July the first *Mariner* was mounted atop an Atlas booster and fired from Cape Canaveral's launch pad 12. It promptly did what all of its *Ranger* littermates before it had done: It failed. As with *Rangers* 1 and 2, the *Mariner* flopped because the booster flopped, losing its way above the ocean and pointing its nose down toward the Atlantic instead of up to space. *And* as with the first *Ranger*s, JPL responded quickly, hammering together another *Mariner,* rolling out another Atlas, and resolving to try again. But even with the scientists working so feverishly, it would still be at least another month before *Mariner 2* could get off the ground, and for Jim Burke, this seemed like a good time to get out of town.

Though Pickering, as head of JPL, had to concern himself with all of the lab's projects, Burke, as the *Ranger* project manager, was strictly a moon man, at least until he was assigned to another beat. For the foreseeable future, there would thus be little for Burke to do but knock around his lunar shop, waiting for the chance to fly again. For the month of August, Burke therefore decided, he and his wife would take a long-delayed trip to Europe.

Pickering might have objected to Burke's decision to leave the country, but if he did, he was not inclined to say so. At least some of the other engineers at JPL almost certainly objected—if only because the absence of the project chief when the project itself was in such distress just didn't look right. Burke, however, was a practical man,

one who appreciated that vacation time was rare, and he was not inclined to sacrifice what little he could find merely for the sake of appearances. As the *Ranger* team sat largely idle and the Venus team
prepared to roll *Mariner 2* out to the concrete griddle of launch pad
12, Jim and Lynn Burke flew off to Europe, spending most of the
month of August touring the countryside of Denmark, the cities of
Germany, and the coolly canopied trails of the Black Forest. Toward
the end of the month, the Burkes arrived in London for the last leg of
their trip. On August 27, at 7:53 A.M. Greenwich mean time—just as
they were sitting down to breakfast in their downtown hotel—the
nighttime skies over far-distant Florida lit up, and *Mariner 2,* bound
for Venus, left the ground.

Several hours later, as Burke and his wife were climbing into their
hotel elevator, the white-gloved hand of a bellboy grabbed the door
and stopped it from closing.

"Mr. Burke?" the boy said tentatively.

"That's me," Burke said.

"Telegram for you."

The boy handed Burke the flimsy yellow paper and released the
door. The elevator started to rise, and Burke tore into the envelope
and pulled out the single sheet inside. The message was from Jack
James, the head of JPL's Venus project.

"*Mariner 2* launched successfully," the three-line letter read.
"Midcourse maneuver pending. We're on our way to Venus."

Burke whooped, clapped, hugged his wife, and only then,
showed her the telegram. When she read it, she grinned broadly and
hugged him back. The youngish man operating the elevator stared
correctly ahead and Burke, impulsively, shoved the telegram toward
him. The operator flicked his eyes downward, scanned the three
lines, and nodded politely.

"That's wonderful, sir," he said, without ever looking at Burke.

Half an hour later Burke was sitting on the bed in his room, trying to make a transatlantic phone connection to California, when
there was a knock at the door. His wife answered and stood aside as
a waiter entered with a chilled bucket of champagne and a vase of
flowers on a linen-covered cart. The accompanying card, embossed

with the name of the hotel manager, read simply, "With our compliments—and our congratulations."

• • •

Little more than a week later a well-rested Jim Burke returned to a jubilant JPL. Little more than two months after that, on October 18, 1962—as *Mariner 2* was passing what was roughly the halfway point on its still-smooth, four-month transit to Venus—*Ranger 5* was launched toward the moon's Ocean of Storms. Deep in the spacecraft's power switching and logic unit was a single bolt that was responsible for the single job of crimping a pair of terminals together, providing electricity to most of the instruments aboard the ship. Unknown to anyone on the *Ranger* team, sometime in the weeks leading up to launch, that bolt had worked its way perhaps a quarter of a turn loose. No one would ever discover why—perhaps it was a careless technician who was thinking about quitting time when he should have been thinking about his work; perhaps it was a single bump on a road the spacecraft encountered on its 3,000-mile truck trip from Pasadena to Canaveral. Whatever the cause, the problem was inconspicuous enough that no one was aware it existed at all until an hour after launch, when a horrified Jim Burke looked at the strip charts and telemetry streaming into the control room and saw the spacecraft power beginning to plummet.

Less than nine hours later it winked out completely, and *Ranger 5*, like *Ranger 4* before it, lapsed into electronic unconsciousness. On October 21 the insensible ship—by now somersaulting uncontrollably through space—flew past the moon and entered permanent orbit around the sun. This time, Burke and the other men of the program knew, there'd be hell to pay.

• • •

William Pickering did not know how he was going to fire Jim Burke. Actually, he didn't even know *if* he was going to fire Jim Burke. He wasn't going to call it a firing, that was for sure. Maybe he'd call it a reassignment. Maybe he'd call it a project swap. And maybe—if he was improbably lucky—he wouldn't have to do it at all. A lot depended on the commissions.

If you had asked any of the engineers and other workers at JPL what to expect after the utter failure of *Ranger 5,* none of them would have failed to anticipate a commission. JPL, after all, was equal parts academic, governmental, and military lab, and if the folks who made up those three groups wouldn't be thinking commission at a time when things had gone so dismally wrong, no one would. What no one would have anticipated would have been *three* commissions.

Almost the moment *Ranger 5* was irretrievably lost, Pickering announced that an internal JPL panel would be convened to look into the causes of the failure and the other four failures that had preceded it. Shortly afterwards, NASA headquarters in Washington trumped JPL by announcing a second board of inquiry of its own. After that, Congress raised them both by calling for a third. Of the three, it was NASA's that looked the most menacing.

While congressional hearings might be a high-profile show, they would be little more than that. *Ranger* engineers might be summoned out of their labs, buttoned into coats and ties, and flown off to Washington to explain to lawmakers why their moon flights kept failing, but in the end, the lawmakers knew full well that they didn't understand a thing about how the space scientists did their work, and if they started mucking around with how the project was run, they stood a good chance of making things worse.

Similarly, the JPL investigation promised to have little bite. There were few people more sympathetic to the trials of an engineer than a group of other engineers, and few engineers they'd be more inclined to sympathize with than the affable Jim Burke, Cliff Cummings, Bud Schurmeier, and the other men of *Ranger.* Indeed, one of the key JPL engineers who would be serving on the commission had himself encouraged Burke to take his vacation back in August. A tribunal of fellow scientists like this might be a lot likelier to get to the bottom of what ailed *Ranger* than a tribunal of congressmen, but whether they would be inclined to do anything about it—firing the friends who needed to be fired and disciplining the ones who needed to be disciplined—would be another thing entirely.

The NASA committee was a different matter altogether. The space agency had an entire exploratory program to run—manned and unmanned missions, near-Earth and deep-space missions—and it was not about to tolerate one small project in one branch office jeopardizing the whole operation. *Ranger* had promise and NASA had been backing it for a while now, but if that promise had faded, the agency would not hesitate to cut the program loose. The new panel was headed by Albert Kelley, a onetime Air Force commander, and at JPL, the day the Kelley commission finally convened and began to hold its hearings was a troubling day, indeed.

Like a storm front, the commission gathered first in the east and moved slowly west, stopping to collect affidavits from people at various NASA facilities and various NASA contractors. Kelley looked into the training of the engineers, the quality of the parts they made, the management skills of their supervisors, the incentives provided for good work, the sanctions imposed for bad work, and, finally, the guts of the spacecraft the scientists put together. The commission started its work in late October, conducted its investigation for the remainder of the month, and finally arrived at JPL sometime in November, spending two weeks interviewing every engineer or administrator who had anything to do with *Ranger,* including Pickering and Burke themselves. By the latter part of the month, Kelley and his men closed their files, gathered their notes, and flew back to Washington to write their report.

For several days not a word was heard from the East. Then several more. Finally, on December 7, Pickering received a bound folder and a phone call. Later that day he made a call of his own, to Burke, telling him he needed to see him.

"Coffee?" Pickering asked when Burke arrived in his office, slightly out of breath from hurrying across the JPL campus.

Coffee, Burke figured was a bad sign. He nodded no.

"Cigarette?"

A cigarette was a worse sign. Again Burke nodded no.

Pickering drew a breath. "Jim," he began, "I've got problems."

Burke nodded, smiling sympathetically and even adding a don't-I-know-it eye roll. Pickering went on.

"I've got congressional committees jumping on me, Kelley's people jumping on me. We've all appeared before them and we know damn well what the situation is."

"We do," Burke said.

"And the conclusion is that we've got to do things differently."

"Okay," Burke said hopefully.

"And change our ways."

"Okay."

"And that means changing people." Pickering gestured vaguely toward a document on his desk. "The report's come down. You and Cliff Cummings are going to have to do something else."

Burke, who had been sitting expectantly forward, now slumped back a bit. Pickering went on hurriedly. "That doesn't mean anyone's leaving the lab," he said. "It doesn't even mean anyone's leaving *Ranger*. It just means we'll put you on a part of the project you're more . . . suited to."

Looking away for a moment, Burke collected himself. He and his friend Cummings were to be sacrificed. His friend Schurmeier was to be spared. He looked back at Pickering. "Do you want to hear from me who I think the new project manager ought to be?"

"Sure," Pickering said.

"I think it ought to be Bud Schurmeier."

"Well, you scored," Pickering said. "Because that's who it's going to be."

Jim Burke, the former head of the *Ranger* program, accepted his new position as lieutenant to his former lieutenant without complaint. From now on he would focus exclusively on the narrow problem of designing new connections between the *Ranger* and the Atlas-Agena and within the *Ranger* itself, making sure there would be no more problems like the ones that plagued *Ranger*s 4 and 5. Cummings would leave the program entirely. After claiming five robot spaceships, the *Ranger* program at last claimed its first scientists. Exactly one week later, the overachieving *Mariner 2* sailed grandly by Venus at a relative cloudtop height of just 23,000 miles—coming more than twice as close to the distant planet as the vaunted Russians ever got.

•   •   •

In the fourteen months after Pickering busted Burke down from general of the *Ranger* division to mere field officer, the program reinvented itself. With Schurmeier now driving the team and Pickering driving Schurmeier, the design and manufacture of *Ranger* 6 proceeded with a near-fanatical attention to technological detail. Manufacturing processes that were once considered routine were monitored by squads of quality-control engineers. Spacecraft assembly hangars that had once been considered more than adequately free of contaminants were now stripped bare and scrubbed raw before so much as a single strut or solar panel was brought inside. Design teams tore into the *Ranger* blueprints, disassembling the schematic ship and not putting it back together until they were sure they had a configuration that would allow it to work right. There would be no loosely torqued bolts on *this* spacecraft, no +3 in the programming where there should have been a −3. There wouldn't even be a balsa wood ball anymore. *Ranger* 6 would be a clean, light, lithe ship, one that would be mounted atop the most exhaustively checked Atlas-Agena ever launched from Cape Canaveral, and sent to the moon to do two things only: take pictures and die.

On the morning of January 30, 1964, that combed and scrubbed rocket took off from Cape Canaveral carrying the rebuilt *Ranger* 6 with it. With Schurmeier in the center seat in Hangar AE, the Atlas and Agena—like few Atlases and Agenas before them—performed precisely as they should, carrying the *Ranger* to just the right altitude, pushing it out of Earth orbit at just the right moment, and sending it toward its lunar intercept on just the right trajectory. So precise was that flight path that if the guidance officers had shut down their consoles that day, their spacecraft would still have arrived in the lunar vicinity less than three days later and flown just six hundred miles wide of the moon, a miss distance that could easily be corrected by the barest pulse from the spacecraft's main engine the morning after launch. Indeed, virtually everything about the early part of the flight looked to be perfect, with one tiny exception.

In the middle of the rocket's climb to Earth orbit—the instant after the Atlas booster shut down and the instant before the Agena lit—a curious bit of telemetry had flashed onto the screens in the control room. The television cameras aboard *Ranger 6*, it appeared, had switched inexplicably on, blinked around briefly in the darkness of their Agena shroud, and, finding no moon there, had switched off again, returning to the long sleep they would maintain until the very end of the mission sixty-six hours later. The entire event had lasted barely eight seconds, and the engineers had not given it a whole lot of thought after that. An anomaly during launch might be troubling, but an anomaly that corrected itself—especially during an otherwise flawless mission—was easy to forget about.

Two and a half days later, however, forgetting about it was not so easy. On February 2, 1964, at 1 A.M., as William Pickering sat in the viewing room of JPL's spanking new—though utterly empty—Space Flight Operations Facility with Homer Newell at his side, hundreds of reporters crowding into the nearby Von Karman Auditorium, and the specter of an angry Lyndon Johnson following things from his White House office, word came back that *Ranger 6*, which was less than fifteen minutes away from crashing onto the surface of the moon, had been struck suddenly blind. The anomalous cameras that had fluttered so suspiciously during launch had apparently blacked out entirely.

It was Walt Downhower who brought Pickering the bad news, flatly intoning, "Still no video, still no video" over the loudspeaker in the SFOF. And as Pickering learned it, the press and the president learned it, too. Now, with the spacecraft approaching 6,000 miles per hour, the moon only a thousand miles below it, and Pickering and Newell alone in their viewing room, Downhower spoke up again.

"At this time," he said, "we are still receiving no high-power video."

"I don't believe what I'm hearing," Pickering muttered. Newell nodded. "I don't *believe* it," Pickering said again. Over the P.A., the high-pitched tones of *Ranger 6*'s telemetry—sounding eerily like the wind the spacecraft would be whipping through if there were any wind where it was flying—was easily audible.

In the old SFOF the *Ranger* engineers were working frantically to get the spacecraft's electronic eyes to open. The newly elevated Schurmeier, who had returned from Cape Canaveral shortly after the *Ranger 6* launch, ordered his imaging chief to send a backup signal to the spacecraft, instructing it to override all existing commands and engage its cameras at once. The signal went out.

"Still no full power video," Downhower's voice said.

Schurmeier told the ship to override the override and try again.

"Full power video still negative," Downhower said.

In Von Karman Auditorium the assembled reporters began to glance at one another. At NASA headquarters they did the same. The mission clock read eight minutes to impact; the cameras had already missed at least five precious minutes of photography time. Around the press room eyes began to roll. With six different cameras aboard this ship, and each one mounted on two different assemblies, probability alone made it all but arithmetically impossible that every part of this superredundant system would fail at once. And yet that's just what seemed to have happened. From a JPL team that had already killed five full spacecraft, however, what more could be expected?

"These guys . . . ," one reporter muttered in disgust.

"Unbelievable," another responded.

"Still awaiting full power video," Downhower answered both of them.

A quarter of a million miles away *Ranger 6* accelerated to 6,000 miles per hour and its altitude shrank from a thousand miles to hundreds. In the empty viewing room Pickering saw the mission clock cross below eight minutes, then seven, then six. If the cameras came to life now, Pickering found himself thinking—this *instant*—he could still salvage maybe two thousand pictures. That number, however, was shrinking as fast as the altitude figure, and on the Sea of Tranquillity, the ten-foot *Ranger* would now be almost close enough to the ground to cast a pinpoint shadow. "Five minutes to impact,"

Downhower said. A minute later it was four, then three, then two.

"One minute to impact," Downhower at last said with finality.

In the press rooms on both coasts, the reporters fell silent; in the viewing room at JPL, Newell and Pickering stared at the mission clock; at his microphone Downhower stopped counting. For sixty seconds nothing was heard in either of NASA's press rooms but the plaintive piping of *Ranger 6*'s telemetry. Then, all at once, it stopped. For an instant there was absolute silence in the NASA halls. Then Downhower spoke up one last time.

"We have first indication of impact," he said uninflectedly. "There is no indication of switchover to full power video throughout the terminal event."

In the new SFOF, Pickering cast one last glance at his mission clock. It read 2 days, 17 hours, 35 minutes, and 29 seconds—the length of *Ranger 6*'s entire life, and not a single picture was returned to mark its passing. Suddenly, however, the P.A. system that had been carrying the spacecraft's telemetry emitted a brief but promising burst of static. Pickering turned expectantly toward it. The reporters and controllers did the same. Somewhere high above them, in the electromagnetic eddies flowing through the upper atmosphere, a signal streaming from who-knew-where collided with the one streaming back to NASA and caught a ride aboard it straight down to the very rooms where things had just gone so silent. When it arrived, that silence was broken by a near-surreal woman's voice.

"Spray on Avon cologne mist," the woman said, "and walk in fragrant beauty."

At the communications station in mission control, the radio man leaped for his console and cut the connection, causing things once again to go quiet. At the control consoles around him the *Ranger* team stared openmouthed. In the press rooms the reporters did the same. In his viewing room Pickering could not help appreciating the dark appropriateness of the moment. His horrific day now had a perfect closing coda. The humiliation of his lab was complete.

•   •   •

JPL engineers did not pretend it would be easy to figure out what had claimed *Ranger 6*'s sight, and they were right. For weeks after the death of the spacecraft, they pored over the strip charts and telemetry records of the flight like heart surgeons reading the final EKG tracings of a man who had died on their table, but they could not find a single scribble that would explain the failure. The camera anomaly during launch was the obvious place for them to start their post-mortem, but to the relief of Pickering and the other controllers who had elected not to look into the problem further, that appeared unlikely to yield anything. Spacecraft were complicated electronic organisms, and, like anything else with so elaborate an anatomy, they were subject to the occasional twitch or hiccup. The twitch *Ranger 6*'s cameras had experienced might have been an untimely one, but it was apparently a harmless one.

Or at least that's the way it seemed. After the initial investigation of the flight had hit its technological dead end, however, Alex Bratenahl, an obscure JPL engineer, approached the *Ranger* team and asked if he might have a look at their launch-day films. Strictly speaking, Bratenahl had no business making such a request. His field of expertise was solar and plasma physics, and he did his work largely in the lab's astronomy buildings—a scientific world away from the *Ranger* team. A theoretical researcher like Bratenahl could work at JPL for years and never have to trouble himself with the crude, workbench business of designing or fixing spaceships. Nonetheless, when he asked for the films, he got them, and within a day after first threading them into his projector he had *Ranger*'s problem licked.

When a gigantic machine like an Atlas-Agena flies through the atmosphere at supersonic speeds, it stirs up all manner of acoustic and aerodynamic hell. One of the most dramatic moments occurs in the middle of that ascent, when the rocket shuts down its main engines, drops its first stage, and prepares for the ignition of the second-stage Agena. At that instant, gases and erupting fluids around the missile cause a sudden, flaming shock wave to engulf it, flowing up toward its nose and then back down again. The shroud of fire lasts less than a second, and is thus unable to do any damage to the skin of the ship itself. But beneath the skin, Bratenahl suspected, it could do quite a bit of harm.

Much of what made up the rocket's veil of flame was plasma, a charged gas that, to the uninitiated, looks like any other gas, but to a plasma physicist is a very different chemical beast. If atmospheric conditions are just right, the almost equal number of positively charged ions and negatively charged electrons in the curious vapor give it the ability to conduct electricity as efficiently as a copper cable. If an Atlas rocket climbing toward space was surrounded by a cloud of the sizzling stuff, and if there was even a microscopic break in the metal shell that enclosed the booster, stray plasma could stream inside, filling the payload chamber with a high-energy cloud. Any piece of electronic equipment the rocket was carrying—including, say, a robot spacecraft on its way to the moon—could then easily switch itself on and roast alive in the supercharged mist. The telemetry data indicating that *Ranger 6*'s cameras had awakened during launch was probably accurate, Bratenahl concluded. The signal that followed eight seconds later indicating that they had shut back down, however, was probably false. The video system hadn't simply turned itself off, it had overheated and died.

The next day Bratenahl returned the films to the *Ranger* team. "Insulate your cameras better," he told them simply. "That should solve your problem." The engineers immediately set about doing precisely what Bratenahl suggested.

But just insulating the *Ranger* spacecraft from injury did nothing to insulate the *Ranger* scientists from criticism, and at JPL they expected a whole different kind of firestorm. To their surprise, this time the criticism was muted. Once again Pickering and NASA convened commissions to look into the latest disaster, and once again the commissions grumbled and fumed and cautioned JPL scientists to watch their steps. Oddly, though, the warnings seemed halfhearted, almost as if NASA no longer expected such agency finger-wagging to have any effect. This sense of surrender was far more ominous than the outrage that had followed the loss of *Ranger 5*. Back at NASA headquarters, Pickering feared, the fate of his *Ranger* program was probably hanging by the most tenuous of threads. Not long after, NASA administrator James Webb called to inform him that that thread had snapped.

"Bill," Webb said when Pickering answered the phone in his JPL office, "we need to talk about a mistake I may have made."

"A mistake *you* made?" Pickering asked.

"You bet," Webb said. "I think I may have gone and given you and your *Ranger* boys more than you can handle."

Pickering closed his eyes. This, of course, was how the manuals said it was done. If you merely plan to reprimand a man, you tell him straight out that he screwed up; if you plan to sack him altogether, you extend him the genteel gesture of pretending that the fault lies with you.

"I can't say I agree with you," Pickering said a little hoarsely.

"I didn't expect you to," Webb answered. "But you're a little too close to the program to see things clearly. From here, it seems like you're in over your heads."

"I think we can climb out of it," Pickering answered.

"You've had time to climb out of it and you haven't yet." Webb took a brief pause and finally said what he came to say. "How about if we just let you fellows out of this *Ranger* contract and turn the whole project over to another contractor?"

Pickering heaved a deep sigh. He had feared he was going to hear something like this; actually, he *knew* he was going to hear something like this. So certain was he that when he did hear it, he felt a perverse sense of freedom. The institutional axe he had been fearing since JPL's *Ranger*s started dropping out of the skies had at last fallen. The lab had failed, Pickering himself had failed, and the profound struggle of making believe that all of the recent failures were mere setbacks, mere engineering challenges, was at last over.

Or it should have been over. The problem was, Pickering believed the make-believe. As he had been taught more than twenty years ago in the dry riverbed that JPL once was, and as he himself had taught his team at the sprawling space center the lab had become, the very heart of the missileman's business is failure—repeated, relentless, humiliating failure—until once, just once, you manage to succeed, and when you do, you know forever how the

machine you're trying to invent ought to be built. With his long-en-
dangered *Ranger* now apparently lost, Pickering had nothing further
to lose by explaining this very thing to the very man who had taken
the spacecraft away from him. For several minutes Pickering held
Webb's ear, telling him that love or hate his *Ranger* program, you
had to admit it had been run exactly the way such a program had to
be run. In the course of explaining all this, he realized that despite
himself, he was also pleading for a chance to try just once more.
Pickering's engineers had fired off six *Ranger*s so far, and with everything that had gone wrong with the program, two of them had flown
straight from Pasadena to the moon. The next one, the JPL boss was
certain—*the very next one*—would land there, too, and before it
crashed, this one would do the job it was built to do.

Webb said nothing throughout Pickering's speech, nothing when
he finished, and nothing for a moment or so after. Only the buzz of
the open connection let the JPL chief know the NASA chief was there
at all. Finally, the instant before Pickering could say anything else,
Webb at last spoke.

"One more flight," he said simply. "You've got one more flight."

•   •   •

Whether the scientists at JPL knew of Pickering and Webb's conversation was impossible for Pickering to know. They wouldn't be likely
to ask their boss if he had been forced to plead for their jobs, and he
certainly wouldn't be inclined to tell them if he had. If the question
was going to come up at all, it would likely come up several nights
later, at the Miss Guided Missile dance. The Miss Guided Missile
dance was an annual late-winter tradition at which the JPL engineers
would gather at a local hall, spend the evening dancing and socializ-
ing, and at the end bring a young secretary or administrative assistant up onstage, place an aluminum foil crown on her head, and
declare her queen of the evening. Most years Pickering would have
been just as happy to give the whole affair a pass, but as head of the
lab he had the responsibility of performing the ceremonial crowning.

The night of this year's dance Pickering arrived late, hoped to leave early, and spent most of the time he was there looking less than enthusiastically forward to the little coronation he would have to perform. Nonetheless, at the appointed hour, a fanfare sounded and someone seized the auditorium microphone and called the lab chief to the stage. Pickering smiled wanly and, rehearsing his little piffle of a speech, began making his way to the front of the room.

As he did, he noticed a reserved, though sustained applause breaking out. The clapping was accompanied by something like a cheer, but a low, contained one—less celebratory than respectful. The cheer built slowly until it was closer to a whoop, then a roar, then an almost ecstatic howl, accompanied by a crack of far louder applause. Pickering looked dazedly around him and pushed his way through the crowd as the sound continued to build. When he at last climbed the stage, the ovation—no denying it was an ovation now— crashed around him. Pickering shadowed his eyes with his hand and tried to squint out over the room, but the bright lights—and the light mist—that were suddenly obscuring his vision made it impossible to see. The applause played out for another minute or two, until finally, the man who had sent *Ranger*s 1 through 6 toward the moon and had just bought his team enough time for a *Ranger 7* leaned toward the microphone.

"We're going to fix it," he said simply. "We're going to make it work."

Nobody was thinking about Miss Guided Missile anymore.

•   •   •

William Pickering's last-chance *Ranger* was assembled at JPL in the spring and early summer of 1964. If the engineers were giving much thought to the possibility that they might never build one again, they didn't show it. There were no final names engraved on any final washers, no teary applause when the little craft at last rolled out. Rather, the ship was simply finished up, checked out, and, like the six *Ranger*s that came before it, loaded unceremoniously into an air- conditioned van for the seventy-two-hour trip to Florida.

The target for the spacecraft this time would be the Sea of

Clouds, a vast plain in the moon's southern hemisphere just to the east of the Ocean of Storms. The announced purpose of all of the *Ranger* flights was to look for landing sites for later manned ships, and the Sea of Clouds, which had been troweled almost completely smooth by lunar lava billions of years ago, was a good candidate. In order to ensure that the pictures this and any other *Ranger* spacecraft took survived—assuming one finally took them—the JPL team planned to process the images in several ways.

All of the photographic data that gushed back to Earth in the final minutes of the spacecraft's life would be caught by the twin, eighty-five-foot maws of the Goldstone antennas. The first and best images these signals would produce would be created by channeling the impulses into an ordinary video system. There a cathode-ray tube would convert the signals into electron pulses and fire them in a rapid scan pattern onto a phosphorescent screen. To the unaided eye, all that would be visible on the screen as this took place would be a rapidly moving pinprick of light, looking less like a picture than a single moving pixel. Undetectably, however, that pixel would be flickering and blinking, tattooing out a pattern that directly corresponded to the shades of black, white, and gray that make up a fully developed television image. A 35mm camera would be positioned in front of the screen with its shutter open throughout this process, so that when the scan was done, a pointillist image would have been constructed on one of its frames. The camera would then automatically advance to the next frame and the next and the next until it had captured no fewer than 4,200 images in just thirteen minutes. The film would then be stored lovingly in a refrigerator and flown to Hollywood to be developed in the best movie industry lab the space agency could find.

This processing would take a bit of time, and if the mission succeeded, NASA knew that that would be too *much* time for the press. In order to satisfy the see-it-now needs of reporters, who would expect pictures to be available within hours, the space agency would also create a quicker, somewhat cruder sets of prints by recording *Ranger*'s data transmissions on tape, playing them back through another video system, and capturing these images with another 35mm camera. If an even quicker look was needed, an ordinary Polaroid could snap a picture of either these recorded transmissions or the raw live ones. Both the Polaroids and the backup set of 35mm negatives would then be flown directly to JPL.

For any lay viewer, the difference between these secondary prints and the better, primary ones would be all but nonexistent. To lunar cartographers, however, the infinitesimal data deterioration that occurs whenever a cruder, second-generation image is created could make the difference between spotting a pothole or boulder on the Sea of Clouds and not spotting it, and this, in turn, could make a huge difference to a lunar crew a few years down the line.

That, in any event, was how the system was supposed to work. But with no *Ranger* having had a chance to try it out so far, no one knew if it would indeed operate the way it was supposed to. Pickering was determined that this time the ground-based cameras would have work to do.

Liftoff for *Ranger 7* was set for July 28, 1964, shortly before 10 A.M., JPL time. Jim Burke, who had overseen five liftoffs as the project manager, would work this launch from an ordinary yeoman's seat in JPL's spanking new, at-last up-and-running Space Flight Operations Facility. Pickering would watch from his no-longer-deserted perch in the SFOF viewing area. A few hundred yards away, newsmen and other JPL employees would once again follow things in Von Karman Auditorium. At 9:50 A.M.—precisely when scheduled—they got the first bit of what they came for, when *Ranger 7* successfully left the pad. Less than fifteen minutes later it left the atmosphere, and less than ninety minutes after that it headed out toward the moon.

During the sixty-six hours the spacecraft was in its translunar coast, the press coverage of the flight was uncharacteristically—but perhaps understandably—sparse. From all signs, *Ranger 7* was on target for its lunar impact, but *Ranger 4* had been, too. Its telemetry stream indicated that all of its systems were functioning, but for a while at least, so had *Ranger 5*'s and 6's. The reporters, it was clear, would be willing to give the story a little coverage, but this time *just* a little. Only when the ship arrived above the Sea of Clouds at about

6:25 on the morning of July 31 and prepared to take its pictures, would the oft-fooled press pay it closer attention.

As the sun prepared to rise on the West Coast that Friday morning, the reporters at both JPL and NASA's Washington headquarters were indeed in their press rooms. *Ranger 7* itself, however, was not precisely where it was supposed to be. With only a few thousand miles to cover and just over an hour of flight time to go, the spacecraft was clearly headed for the Sea of Clouds, but a subtle drift in its flight path had nudged it a bit off course, causing it to approach its impact site at a slight angle, one that sharpened shadows and distorted perspectives slightly. The mission directors were fully aware that this might happen and had built an opportunity for an eleventh-hour trajectory adjustment into the flight plan. At shortly after 5 A.M., that eleventh hour arrived, and Bud Schurmeier now had to decide whether to fire his ship's thrusters and change its orientation slightly or simply let it fly on.

At his console in the SFOF, Schurmeier huddled with Gerard Kuiper, the head of the experiment team that would be analyzing *Ranger 7*'s images. Kuiper, he knew, had been huddling with his own scientists, and he had a pretty good idea what message he came bearing.

"We've been talking it over in the back room," Kuiper said, "and we really think we need to nudge this ship."

"Which means disturbing its guidance system," Schurmeier said.

"I know," Kuiper answered.

"And firing up its thrusters."

"Mm-hmm."

"And you think it's worth it?"

"I do," Kuiper said.

Schurmeier nodded thoughtfully. "I don't," he said. "I don't want to touch this spacecraft unless it's absolutely necessary."

"If we don't get the right kind of pictures," Kuiper objected, "we miss half the point of the mission."

"If we don't get any," Schurmeier answered, "we miss it all. We'll go with the trajectory we've got."

Kuiper nodded once and Schurmeier keyed open his microphone. "Terminal maneuver will be a no go," he said with finality. In the viewing room Pickering smiled approvingly; at his console on the SFOF floor, Burke agreed. On the wall above them, the mission clock read barely one hour to impact.

For the next forty-plus minutes *Ranger 7* would do nothing note-worthy—not until precisely eighteen minutes before impact when the on-board timer would order the cameras to begin charging them-selves up. Three minutes later, just as they had on *Ranger 6,* they would be instructed to open their eyes and send their pictures home. When the mission clock read thirty minutes to impact, the men in the SFOF began nervously studying the trajectory plots. When it read twenty minutes, the ones who had been milling about returned to their seats. When it read nineteen minutes to impact, they leaned to-ward their consoles. At the eighteen-minute mark, Schurmeier squinted at his screen, drew a breath, and then released it in a great gust as the tiny data point on the monitor indicated that the cameras were indeed receiving juice.

"Cameras in full power mode," he announced. The men at the consoles made celebratory fists; the men in the viewing room flashed them a thumbs-up.

In Von Karman Auditorium, George Nichols, a *Ranger* engineer, would be performing the public address job Walt Downhower had handled on *Ranger 6.* Listening through his headset, he heard Schurmeier's call and repeated it to the room. "Ranger Control re-ports that cameras are in full power mode," he said. The press im-mediately broke into applause and Nichols quickly hushed them. "We point out that having the TV system in full power mode is not proof that the cameras are operating," he called out. "That will not come for another three minutes."

For the people in the SFOF and the two NASA press rooms, that time would play out slowly. Before the necessary 180 seconds elapsed, the ship would have to free-fall at least three hundred more miles, plummeting from 1,800 miles above the Sea of Clouds to just over 1,500. The first minute ticked away, then the next, then the next. Then, at almost precisely the instant the mission clock turned over from 15:00 to 14:59 and the spacecraft plunged through the 1,520-mile altitude mark, the antennas in the Mojave Desert twitched to life.

"Video signal received!" a Goldstone technician shouted into his headset to JPL. "Video signal received! It looks strong and it looks clean."

Schurmeier grinned broadly. "Strong and clean video being received by Goldstone!" he repeated to the SFOF.

In Von Karman, Nichols made the call, too. "At this point," he said evenly, "a video signal is being received and is reported to be strong and clear." A roar went up in the room, and Nichols again tried to override it. "This-does-not-mean-a-picture-is-being-received," he shouted in a hoarse staccato. "It will take some analysis and a check of the recorders to determine that."

Nichols was right, of course. If the signal flowing from the spacecraft was weak or garbled, the desert antennas would be receiving little more than cosmic nonsense. But even as Nichols was making his cautionary announcement, equipment began to stir deep in a Goldstone lab. On a bank of black-and-white television monitors, tiny, phosphorescent fireflies had appeared and begun tracing a crazy path across the glass, and a series of cameras mounted in front of them began to snap and snap and snap. At one monitor a Polaroid camera clicked, whirred, and spit out a single card of photographic paper. An engineer standing by instantly grabbed it and ripped the backing away. In front of him he saw what the spacecraft's cameras had seen, and what the spacecraft's cameras had seen were hills and rills, craters and gullies, canyons and plains and lunar seas. What the cameras had seen was the moon.

"We've got pictures!" the technician called out.

"Pictures are in!" a Goldstone communicator relayed to Schurmeier.

"Goldstone has pictures!" Schurmeier repeated to the SFOF.

In Von Karman, Nichols braced. "Preliminary analysis of the signal," he said, "is that we are seeing pictures." This time, he knew, there would be no silencing the reporters' cheers. All over the twin NASA press rooms, shouts went up, hugs were exchanged, notebooks were tossed in the air. In the SFOF viewing room, Pickering

and the other administrators reacted with similar—if more reserved—glee, rising almost as one, shaking hands, flinging arms
around shoulders, and applauding down at the mission control team
working below them. The team members themselves looked over
their shoulders at their appreciative superiors and broke into huge
and helpless smiles.

For the next fourteen minutes the 807-pound *Ranger 7* continued to fall and its bank of six cameras continued to fire. Dozens,
then hundreds, then thousands of electronic images deluged the
thirsty Mojave antennas. As film spun through the 35mm cameras
and picture after picture dropped from the Polaroids, the whistle of
the accelerating spacecraft's trajectory seemed to grow louder and
louder in the headsets of the controllers. The ship plunged from an
altitude of 1,000 miles, down to 500, and then on down to just a few
dozen as its speed climbed and climbed and climbed. From the surface of the moon, what had appeared only moments ago to be a single, stationary star would now appear to be a moving, glinting,
growing mass. Finally, at exactly 6:40 A.M. Pasadena time, after returning 4,316 pictures, *Ranger 7* completed its plunge, slammed into
the ancient surface of the Sea of Clouds, and, as the physicists predicted it would, liquefied, vaporized, and ceased to exist.

More than 240,000 miles away, a team of astronomers in a
British observatory were certain they saw the flash.

•   •   •

Less than three hours after the spacecraft that had been known as
Ranger 7 passed into history, William Pickering retreated to his office and invited Jim Burke, Gerard Kuiper, Bud Schurmeier, and a
few other *Ranger* scientists to join him there. A few minutes after
they arrived, a delivery boy showed up with sandwiches and sodas.
A few minutes after that, there was another knock at the door.

Pickering rose and opened it, and a young technician from the
JPL photo lab entered carrying a portfolio-sized manila envelope.
The technician handed the envelope wordlessly to the JPL chief, who
nodded his thanks and, with a smile, beckoned the boy to leave. Af-

ter he had gone, Pickering opened the little metal clasp at the top of the envelope, lifted the flap, and slipped his hand inside, sliding it beneath a stack of slightly warm photographic paper. Easing the stack out with the care of a waiter maneuvering a heavy tray, he nodded to the men in the room, who swept the cups and food wrappers off the table where they had been eating. Carefully then, Pickering laid the moon before them.

The senior members of the *Ranger* team looked down at the images on the table in something close to utter silence. Someone might have murmured "astounding"; someone else might have murmured "remarkable." Whatever was or wasn't said, the pictures their ship had brought them were extraordinary. There were great plains and small flatlands, jagged peaks and smooth hills. Over here was a chasm, over there was a crater, over there was a boulder as small as a car. The scientists had looked at thousands upon thousands of lunar images before, but all had been taken by telescopes anchored on Earth. These pictures were taken by the first American camera to cross the translunar distance and do its work up close. Slowly, slowly the scientists flipped through the stack of images, and then when they were done, they flipped through them again.

The quiet scene in Pickering's office took an hour or so to play out and was nothing like the bedlam that had broken out just a little earlier in Von Karman Auditorium, when the same men made their first appearance in front of the press after the spacecraft's impact. Even before the scientists arrived, the newsmen had risen and offered an anticipatory round of applause to their tiny TV images when the SFOF camera caught them in its sights and flashed them onto the press room monitor. When the scientists arrived in Von Karman, the applause turned into a cheer. The moment the *Ranger* bosses mounted the stage, the already boisterous crowd hurled questions at them.

"How good did the Goldstone people say the pictures are?" someone asked.

"Several times better than any pictures of the moon we've ever seen before," Pickering said judiciously.

"Have they seen anything unexpected?" someone pressed.

"If you mean were there any little green men," Pickering said, "the answer is no." The room erupted in laughter, though at this point Pickering suspected he could have read aloud from a grocery list and this crowd would still have responded happily.

"What do you think about the future of JPL now that this mission has succeeded?" a voice called out.

This time it was Pickering's turn to laugh. "I think," he said, "it's improved."

Later that day a phone call came in to Pickering from Lyndon Johnson. The president congratulated the *Ranger* team on its work, inquired about the newly returned pictures, and allowed as how he wouldn't mind seeing a few of them himself. Lyndon Johnson allowing that he wouldn't mind seeing something meant that he jolly well expected to see it, and later that night a somewhat dazed Pickering found himself on a red-eye to Washington. By eight the next morning, he was standing in the Cabinet Room in the White House, with the president, a handful of senators, and Jim Webb, the beaming NASA administrator. With the help of an easel, a pointer, and blowups of the same pictures Pickering himself had seen the previous afternoon, the JPL chief showed the president the curious place his space agency's spaceship had visited the day before. Johnson listened attentively, clucked over the pictures, and once again offered Pickering his congratulations. An hour later Pickering was on a plane headed west. Six hours later he was back in his lab, as if his trip to the White House hadn't taken place at all. For all he knew, in the president's mind, it never had.

Seven months later, in February 1965, *Ranger 8* flew flawlessly to the moon, sending back thousands of pictures of its own before annihilating itself against the ancient face of the Sea of Tranquillity. A month after that *Ranger 9* followed, successfully self-destructing in the Alphonsus crater, after returning a third album of images. Following that, Pickering knew, no more *Ranger*s would be flying. Now that NASA had proven it could get where it wanted to go, it had far bigger plans to pursue. There were the *Surveyor* ships that would soft-land on the moon in a single, functioning piece; there were the *Viking* probes that would accomplish the same feat on Mars. There were the *Mariner*s to Venus, the *Pioneer*s to the sun, and the other, as yet unnamed ships that might go as far out as Jupiter. And then, of course, there were the grand *Apollo* skyliners that would take not hardware, but humans to the same moon *Ranger* had so recently claimed.

With such magnificent ships to build, the agency didn't have much time for little spacecraft able to do no more than fly out to the moon and belly flop in its soil. *Ranger* may have been the first probe to make such a deep-space trip but it wouldn't be the last, and before the final, vaporous remains of *Ranger 9* had even had a chance to scatter in the nonexistent lunar winds, Pickering knew it was time to turn his mind to other things. It would be awhile, he realized, before he'd be getting another call from the White House.

Early 1966

Jim Burke was finding it hard to believe what the numbers coming out of the computer were telling him. As an engineer Burke had long since learned to trust the numbers he worked with every day, even if what they suggested and what common sense suggested were two different things. Today, however, the numbers were suggesting something fantastic.

In recent months Burke had grown increasingly familiar with fantastic numbers. It had been about a year since the end of the *Ranger* program, which meant it had been about a year since he had started working for Homer Joe Stewart. Homer Joe Stewart, as anyone at JPL could tell you, was a bit of a visionary, and thus it was fitting that when he got his own JPL section to head, it was an unconventional one. Stewart's group wasn't involved in something as mundane as developing hardware or flight plans for current missions, but cooking up schemes for future ones. If there was a planet no one had even considered visiting yet, it was Stewart's group that concocted a way to get there. If there was a spacecraft no one had even dreamed of building, it was Stewart's group that would dream it up. For an engineer like Burke, who was accustomed to assembling here-and-now machines, this kind of blue-sky thinking was a new thing. Gradually, however, he'd gotten used to the way Stewart's department did things, knowing that if the ideas were good enough and the ships imaginative enough, at least a few of the whimsical flights might actually get flown. The data coming from the computer today was the stuff of whimsy, indeed.

Burke had been called down to the computer room by Jim Long, a more junior member of the department who had been exploring the idea of missions to the solar system's outer planets. Long, it was evident, loved playing about with long-distance flights like these. The billions of miles the ships would have to cover and the dozen or so years it would take them to make their journeys made *Ranger* and *Mariner* missions seem like day trips, and Long used to spend untold hours hunched over his solar system maps, happily plotting the imagined ships' improbable trajectories. When Burke arrived in the computer room a few minutes ago, Long looked even more pleased than usual, and before the senior man could say anything the junior man thrust a sheaf of papers at him.

"Look at this," Long said. Burke started to respond, but Long cut him off. "Just look at it."

Burke glanced at the page and saw that it was dense with numbers and planet names. He looked a little closer and saw that what he was seeing were orbital calculations, but not present-day ones; these were projections for the orbits of the planets more than a decade in the future. He looked closer still, and the full impact of what he was seeing hit him. He whistled to himself softly and turned to Long with a smile.

"When will this happen?" Burke asked.

"Nineteen seventy-nine," Long answered.

"You're sure of this?"

"Completely."

Burke smiled again. What Long was telling him was remarkable. With nine planets in the solar system and each orbiting the sun at different radii and different speeds, the local family of worlds looked, to the lay eye, like nothing more orderly than a nine-ball roulette wheel. But even out of chaos can come order, and the solar system was no exception.

Every 176 years, random orbital motion caused the four large, outer planets—Jupiter, Saturn, Uranus, and Neptune—to organize themselves in a near-perfect line. The tidiness wouldn't last long; look away for too long and when you looked back, things would have already started to deteriorate again. But if you acted fast—if you acted with foresight—you could launch a spacecraft from Earth that would reach the outer system just when the four gas giants had begun to form ranks. With the right navigational plan, you could then tour them all, using Jupiter's gravitational energy to fling you on to Saturn, Saturn's to get you to Uranus, and Uranus's to get you to Neptune. Neptune would then toss you out of the solar system altogether, sending your spacecraft not just into the void between the planets, but into the void between the stars. The idea wasn't a completely new one; JPL mission planners had been talking about the possibility of just such an orbital alignment for a long time. Now, however, it looked as if it was certain to happen—and happen soon.

Burke looked up at Long again.

"When do you think we'd need to be ready to launch?" he asked.

"Nineteen seventy-seven," Long said.

"Then we need to start planning now."

2

Green Glass and White Rocks

Thursday, July 29, 1971, 9 P.M.

Dave Scott was two-thirds of the way to the moon when he at last caught the smell of pork and scalloped potatoes cooking in his command module. The *Apollo 15* crew had smelled a lot of smells in the eighty or so hours they'd been away from Earth—the new rubber of their spacesuit liners, the fresh canvas of their couches, the vaguely acrid scent of oxygen that had traveled through hundreds of feet of plastic and metal tubing before at last flowing into their tiny cockpit—but none of them had been as welcome as the pork and potatoes they were smelling now.

To be sure, it was something of a stretch to say the pork and scalloped potatoes were cooking at all. Cooking required an oven, and an oven required room, and room was not something three men could hope to find in their cramped little pod of an *Apollo* spacecraft. In another year or two NASA would be launching its much talked-about Skylab space station. While the huge orbiting liner wouldn't travel very far from Earth—permanently circling the planet just 235 miles up—it would keep its lucky crew of three in the kind of comfort only a big ship with lots of interior acreage could provide. You could bet Skylab would have an oven on board, and when it came time for dinner on *that* ship you'd get genuine smells of genuine food from one end of the forty-five-foot craft to the other.

Aboard *Apollo* things were different. All of the food for a two-week lunar journey was stored in a foot locker–sized larder tucked away at the foot of the command module. The 126 meals—three a day for three men for fourteen days—were freeze-dried, shrink-wrapped, and then color-coded red, white, or blue, indicating the commander, command module pilot, or lunar module pilot. When mealtime rolled around, the crew members would simply scoop out a few packets, snip the ends off with a pair of shears, and then inject hot water into the opening with a hose and gun assembly stashed in the lower equipment bay. With much of the food—processed and puréed to an all but unidentifiable paste—it wasn't always easy to taste what you were eating, much less smell it. But for some reason, pork and scalloped potatoes were a different story. Hydrate the meal with just enough water, and the fragrance coming out of your packet—if not the look of the food inside it—could almost make you believe you were eating an actual Earth meal in an actual Earth kitchen.

Dave Scott had been looking forward to pork and potatoes night for a while now, and as the menu rotation would have it, it fell on the best of all possible evenings. In just eighteen hours—or what would be mid-afternoon on Friday if the crew was still in Houston—he and Jim Irwin would separate their spindly lunar excursion module from their cone-shaped command module, fire their descent engine, and head down to the surface of the moon. A few hours after that the LEM would touch down on a prairie known as the Hadley Plain, hard by the toothy Apennine Mountains, just near the spot where the waters of the Sea of Serenity would overlap the waters of the Sea of Rains if the Sea of Serenity and the Sea of Rains, in fact, had any water. It was a spot Scott had lobbied hard to visit, and tomorrow at this time he was going to arrive there. For a man hoping to study lunar geology, there was no better place to be.

For all their trailblazing achievements, *Apollos 11* and *12*—the first two lunar landing missions—had played it relatively safe. The target for those missions, both of which flew in 1969, had been the Sea of Tranquillity and the Ocean of Storms, and if you were looking for a good, safe spot to set down a lunar lander, you couldn't pick better runways. For one thing, both places were utterly flat. It wasn't for nothing that ancient astronomers had mistaken the vast, unbroken stretches of extraterrestrial terrain that mottled the moon for seas, and for acre after acre, mile after mile, they rolled on as huge and featurelessly as the bodies of water for which they had been misnamed.

For another thing, both places were well mapped. In 1965, *Ranger 8* had ended its brief life in the Sea of Tranquillity, transmitting more than seven thousand pictures of the surrounding landscape before destroying itself in the ancient hardscrabble. In 1966 and 1967, *Surveyor*s 1 and 3 soft-landed in the Ocean of Storms, touching gently down on a plume of rocket exhaust and returning nearly 17,000 pictures of the land around them. Shortly after, *Surveyor 5* followed, returning to the Sea of Tranquillity, soft-landing not far from where *Ranger 8* had crash-landed, and broadcasting home more than 18,000 photographs to supplement the ones already in hand. In addition to all that reconnaissance, NASA had dispatched a fleet of five *Lunar Orbiter* ships to circle the moon between 1966 and 1967, conducting further photo surveys of all of the largest lunar seas—Tranquillity and the Ocean of Storms among them. After so much high- and low-altitude surveying, there was barely a rock or divot in the surface of either of the two plains that hadn't had its picture taken over the last few years. While NASA flight planners couldn't guarantee that an errant crater or boulder field wouldn't still pop up to surprise an approaching astronaut crew, they were pretty sure that they were now utterly familiar with the lay of the lunar land.

Sending astronauts to such safe and nondescript places, however, came at a price. From the moment modern telescopes and robotic spacecraft confirmed that the moon did not appear to have ever held so much as a cupful of water—or, for that matter, a lungful of air—it seemed clear that one thing its seas did hold was lava. With no wind or rain to sculpt the land, there was only one other force besides water that could trowel the lunar seas down as flat as they were, and that was molten rock pouring over the surface from magma chambers far below.

The problem with studying land like this is that if you're looking to peek back into lunar history, a lava plain can take you only so far. Geologists seeking to determine the age of surface formations lived by a simple, sacred doctrine known as the Law of Superposition. For all of the arcane chemistry and complex mineralogy that is part of the geological art, the Law of Superposition was a wonderfully straightforward idea. The deeper a geographic feature is buried, the law decreed, the older it is; the shallower it lies, the younger it must be. A geologist looking at a photograph of a rock lying on the lip of a crater, which was itself dug out of the foot of a hill, knew instantly and intuitively that the hill was pushed up first, the crater blasted out of it later, and the rock—through some other dynamic event—deposited on it last.

Applying this same reasoning to lunar seas, the scientists realized that these huge flatlands—which covered everything around them and yet were themselves marred by almost nothing at all—had to be among the youngest features on the by-no-means young moon. If you wanted relatively recent geology, the seas were a perfectly adequate destination. But if you wanted to collect the true keystones to the lunar past, you had to head for the lunar hills.

In 1970, *Apollo 13* had tried a landing in the moon's highlands, setting out for the hills of the Fra Mauro formation in the moon's southwest quadrant. After a translunar explosion blew the ship and the mission all to hell—forcing commander Jim Lovell and his rookie crew to turn around and fly their mortally wounded spacecraft back home—*Apollo 14* aimed for, and this time reached, the same site. Now, *Apollo 15* was going to raise the topographic ante, bypassing the plains and the hills, and heading straight for the base of the Apennine Mountains, three-mile-high peaks that were among the starkest features anywhere on the moon.

There was no telling exactly how a mountain range as dramatic as the Apennines formed, but geologists had a pretty good idea. Long ago, they hypothesized, the moon was a fairly undifferentiated place, with perhaps a few wrinkly hills breaking up its otherwise egg-smooth surface, but not much else. Now and then, however, a mammoth boulder cannonballing through space would take aim at the airless—and thus defenseless—moon and collide with it broadside. Vaporizing on impact, a meteorite this big wouldn't leave a simple crater, but instead would gouge out a valley that might measure hundreds of miles across. The injury the projectile inflicted would not be just wide but deep, causing a great hemorrhage of the moon's molten lifeblood to flow out over the surface. The bleeding would continue for tens of millions of years before it was stanched, and when it finally was, it would leave behind a smooth, new lunar sea across most of the valley the collision had created.

Around the edge of the newly formed basin, things would not remain so flat. The land the meteorite excavated when it hit would have to go somewhere, and in general, where it would go would be out and up, crumpling toward the sky in huge, instant mountain ranges that would surround the sea. Since the new peaks would be spared the deluge of lava that flowed into the valley below them, their flanks would still be made almost entirely of ancient crustal material. A crew of astronauts who put their spacecraft down near the mountains would thus merely have to look for a little high-lying rubble and they would likely have themselves an authentic bit of the original moon.

It was precisely this kind of debris the *Apollo 15* crew would be hoping to find when they landed at the foot of the Apennines, and the material they would be keeping a special eye out for would be something called anorthosite, a bright, white crystalline material made up principally of the mineral plagioclase, which was itself made up of aluminum and calcium. Scott and Irwin had seen plenty of anorthosite on geology treks they'd taken into California's San Gabriel Mountains over the course of the last year; now, they'd be looking for the same common debris on the surface of an entirely uncommon world. If they were able to bring a little bit home, and if the age of the sample was older than the age of the lunar sea itself, they would both prove the impact theory of basin formation and confirm the age of the moon as a whole.

As a novice lunar geologist, Scott looked forward to digging through the soil to hunt for the ancient samples. As a veteran astronaut and test pilot, he looked even more forward to the business of getting there in the first place. Steering his little, foil-skinned LEM through the spires of the Apennines and setting it down at the base of the mountains would be a nifty piece of piloting—a *sporty* piece of piloting, as Scott liked to say—and he was only too pleased to try his hand at it.

Tonight, at 9 P.M., he was just eighteen hours away from getting his chance. Floating above his canvas seat on the left-hand side of the spacecraft, he scooped a spoonful of pork and scalloped potatoes from his plastic food pouch and chewed slowly and silently. In the center seat, Al Worden—the command module pilot, who would stay behind the wheel of the lunar orbiter while his crew mates were down on the surface—did the same. Off to the far right, Irwin mirrored them both.

Scott turned idly to his left and happened to look out the window on the port side of the cabin at just the moment the wasted landscape of the Sea of Serenity rolled by sixty miles below. Next to it lay the vast, gray stain that was the Sea of Rains, or Mare Imbrium, as the scientists both on the ground and in the spacecraft less colloquially called it. Somewhere on Imbrium was the spot where Scott and Irwin would be walking in less than a single Earth day. Scott scanned the terrain and broke into a broad grin.

"Hey, Jim," he called to Irwin. "You know what I'm doing tomorrow?"

"Can't guess," Irwin answered, with a smile of his own.

"I'm gonna go shoot a landing"—Scott looked out his window, found his spot, and pointed—"right there. I'm gonna put that baby right down there."

"That's right," Irwin said, a little less certainly.

"Yes, sir," Worden agreed.

On the starboard side of the cockpit, Irwin shifted his body slightly, peered out his own window, and watched the Mare Imbrium landing site Scott had just spotted slide further into view. Unlike his commander, who had flown in both the *Gemini* and *Apollo* programs, Irwin was new to the outrageous business of space travel. And unlike Worden, who was also a novice, he would not be spending the entirety of his maiden flight in the relative safety of the command module. Rather, he would be relying on his rookie skills—and his rookie nerves—to see him all the way down to the surface of the moon.

"You know what I want to do first thing when we get back home?" Irwin said after a moment's reflection.

"What's that?" Scott asked.

"Have a beautiful night in Tahiti."

"You're on, buddy," Worden said. "You're on."

"You bet," Scott agreed.

"No, really," Irwin persisted, all playfulness gone. "I'm not kidding. We really ought to think about it."

"Absolutely," Scott agreed. "We'll book a big airliner."

Irwin looked around himself at the little nutshell of a command module that was keeping him and his crew mates alive; then he cast his eye out his forward window, where he could just see the roof bolts of the fragile lunar module that would soon be transporting his mortal soul to the foot of the Apennine Mountains and back again. A veteran like Scott might never quite get his fill of flying, but after this week a first-timer like Irwin would have had enough to last him quite a while.

"Without the aviation," he said flatly to Scott. "Without the aviation."

July 30, 1971, 5 P.M.

The *Apollo 15* moonwalkers were under no illusions that flying their lunar module down to the surface of the moon would be a simple matter. What made it even harder was the fact that they'd be flying it blind.

Earlier today, at just 3:00 P.M. Houston time, Scott and Irwin had climbed into the LEM, sealed their roof hatch, and cast off into orbit. When they did, they were moving at about 5,000 miles per hour—a velocity that would have kept them in lunar orbit indefinitely. In order to descend to the surface, they would have to reduce that speed by more than half; in order to do that, they would have to position their LEM so that its descent nozzle faced forward, then hit the ignition and let the chemical fire that poured from the engine bell apply a propulsive brake. The problem with this simple maneuver was that since the LEM was designed to land in a standing position, its descent nozzle was built into its underside. Pointing the engine forward thus required rotating the ship 90 degrees backward, and that left the astronauts lying flat on their backs—flying feet first, head last, eyes looking up at the stars.

For almost any pilot, this was an awkward—not to mention undignified—way to fly, yet for most of the sixty-mile descent to the moon, that was just what the mission plan demanded. It was only when the crew was 7,000 feet above the surface and three minutes from touching down that the on-board computer would at last order the clusters of thrusters arrayed around the ship to bloom to life, rolling the LEM to an upright position. The commander, with his feet back beneath him and the lunar landscape flooding his window, would then grab his maneuvering stick, scan the terrain for familiar surface features, and at last begin the high-wire flying he came here to perform.

Lunar module crews—who had spent months studying terrain maps before setting out for the moon—typically had a lot of landmarks to help them steer their ships in, and Scott and Irwin were no exceptions. In addition to the mammoth Apennine Mountains, which would serve as their first and most prominent harbor marker, they would be looking for a depression known as Spur crater, a deep abyss that marked the edge of their prime landing site. Also helping to frame the target area would be Hadley Tower, a sheer 11,000-foot mountain at the edge of the Apennines; the North Complex, a small cluster of prominent craters; and Hadley Rille, a sinuous lava channel snaking through the surrounding Hadley Valley.

Along with these friendly landmarks, there was also one that wasn't so friendly. On the other side of Hadley Rille was a single, low hill easily visible from high above. If Scott steered his ship too close to that unmistakable landscape bump, he would know that he had overshot his prime target, was drifting well beyond the area NASA considered safe, and was in danger of putting the LEM down on a deadly, boulder-strewn incline. Scott, recognizing the dark significance of the little peak, had named it Bennett Hill, after Floyd

Bennett, the NASA trajectory planner who had designed this dare-
devil approach in the first place. The name Scott had chosen for his
lunar module itself was equally apt: *Falcon,* he had decided to chris-
ten the ship, in a nod to the acrobatic flying it would have to perform
in order to get through this expedition intact.

The descent to the surface began not long after Scott and Irwin
climbed into the lunar module and drifted away from the command
module, when they at last instructed their computer to instruct their
descent engine to begin firing. For an instant they felt nothing at all,
then a low-register rumble shook the cockpit, while outside, an utterly silent column of flame erupted from the bottom of the ship. Almost undetectably, Scott and Irwin noticed a faint sense of pressure
as the LEM began to slow and the barest ghost of gravity appeared.
On their instrument panel the numbers on the velocity indicator began to fall and the trajectory readouts began to steepen. Nearby, the
altitude gauge, which had been holding steady at 60 nautical miles,
began to slip, too.

For more than seven minutes, the moonwalkers sailed silently
along, riding the quiet fire issuing from their ship, as they soared
over the Sea of Serenity, Mare Imbrium, and the occasional crater or
wrinkly rill that here and there disfigured both. Finally, when the *Falcon* had descended to the prescribed 7,000-foot altitude, its thrusters
puffed to life and it swung itself slowly upright. Scott looked immediately out his left-hand window, Irwin looked out his right, and
both men went wide-eyed.

Surrounding the ship wherever the astronauts looked—above
them, below them, across the entire field of vision their forward-
facing windows allowed them—stood the sheer, gray-white walls of
the Apennine Mountains. For most of the last twenty-four hours the
Apollo 15 astronauts had been seeing the moon from a nosebleed-
high five dozen miles that reduced even the grandest features on the
lunar surface to little more than bumps. Now, Scott and Irwin were
just over a mile above ground, flying through a mountain range
whose loftiest peaks were nearly twice the altitude of the ship itself.

Scott looked out his window and took in the mammoth canyon
walls that hulked all around him. Though the moment seemed to call

for some acknowledgment, he decided to limit his comments to pretty much what he had been trained to say at a time like this—which was pretty much nothing at all. The final descent into the Hadley region would require utter concentration, and if there was any talk going on inside the cockpit at all, it would come from Irwin, whose responsibility would be to keep an eye on his instrument panel and read out the spacecraft's ever-changing speed, altitude, and angle of descent. Scott, in turn, would keep his gaze fixed out the window, working the ship's maneuvering stick in response to Irwin's call. Irwin let a few long moments elapse as the ship continued its steep descent, gave his gauges a fast scan, and announced his first reading.

"Five thousand feet," he said uninflectedly, "thirty-nine degrees."

Scott nodded to himself and tweaked his throttle higher. The pitch of the engine picked up a bit and vibrated through the cockpit. Irwin let nine more seconds elapse.

"Four thousand feet," he said, "forty degrees."

Scott nodded again, tweaked again, and listened as his engine climbed half an octave again. He kept his eyes fixed out the window and worked his stick by feel alone.

"Three thousand feet, fifty-two degrees," Irwin said.

With his throttle held steady, Scott squinted out over the landscape below, scanning for familiar features. In the distance he thought he could just make out the serpentine line of Hadley Rille and the dark well of Spur crater. Ten more seconds elapsed.

"Two thousand feet, forty-two degrees."

Scott now saw a few more craters freckling the landscape. All of them were familiar to him, just where the mapmakers said they would be. Nearby, he also spotted a patch of smooth, invitingly unbroken ground. He smiled slightly.

"Okay," he said, for the benefit of both Irwin and Mission Control, "I got a good spot."

"Good," Irwin answered. "Forty-two degrees."

Scott throttled his engine to slow his descent still further and the *Falcon* went into what briefly felt like a stationary hover. He engaged his array of sixteen thrusters and the ship edged forward toward the target spot on the surface. Over the next minute the ship descended to 1,000 feet, then 500, then just 100. Finally, sixty seconds before scheduled touchdown, it had slowed to a speed of just three feet per second and an altitude of just sixty feet, and the tongue of flame that was holding the *Falcon* up at last began to graze the ground. When it did, dust devils began to rise and churn, swirling wildly outside the astronauts' windows.

"Okay," Scott said evenly, "I've got some dust."

Some dust quickly turned into a lot of dust, and a lot of dust quickly became a whiteout. On the other side of his triangular porthole, Scott could now see nothing at all. Holding tight to his throttle, he eased the ship lower and lower, as the engine's vibration grew louder and louder. At last, when the LEM was just three feet above the surface, probes extending from the bottom of its foot pads made contact with the ground and a sea blue indicator lit up on the instrument panel.

"Contact!" Irwin called.

Scott immediately reached forward and hit his engine stop button. Instantly, the pillar of fire beneath the ship vanished and the *Falcon* free-fell the last thirty-six inches to the surface, hitting with a deadweight thud that caused every piece of equipment in the cockpit to rattle in its bracket.

"Bam!" Irwin called out.

Bam, Scott echoed in his head. Without a word he reached forward and threw a bristle of switches, disarming his engine and reconfiguring the controls of his now stationary ship. Finally, he drew a breath, keyed open his mike, and called out to his home planet 250,000 miles away.

"Okay, Houston," he said with satisfaction, "the *Falcon* is on the plain at Hadley." And so it was.

August 1, 1971, 10 A.M.

Dave Scott and Jim Irwin knew enough not to say too much when they found their $400 million rock. It was hard to determine exactly what the little lump of lunar rubble was worth, of course. But since it had cost $400 million to build and fly *Apollo 15,* and since this was precisely the rock they had traveled across the void of translunar space to find, $400 million seemed like a pretty fair figure. In any event, it didn't pay to talk too much about it.

It was Buzz Aldrin who had first learned the lesson of talking too much on a lunar geology expedition, and he learned it the hard way. Buzz had preceded Scott and Irwin here almost two years ago to the day, when he and Neil Armstrong put their *Apollo 11* LEM down in the Sea of Tranquillity, 423 miles to the southeast as the lunar crow flies. Like so many of the pilots who were chosen for the astronaut program before him, Aldrin got his education at West Point, studying history, science, and soldiering in pretty much equal measure. Like not so many others, he decided to continue his schooling after that, earning a Ph.D. from MIT in the brain-numbing business of astronautics and orbital physics. A man who could fly airplanes, master the sciences, and pile up degrees like so much kindling was just the kind of man NASA was looking for, and it was no surprise that when it came time to pick a crew for the first lunar landing, Dr. Buzz Aldrin earned himself a seat.

Buzz took enthusiastically to his astronaut training—no more so than when that training included geology. For nearly half a year before *Apollo 11* took off, Armstrong and Aldrin made regular trips into the California badlands to learn the spelunking skills they would need when they at last reached the surface of the moon. More than a few of the scientists who came along with them on these expeditions had serious questions about the value of the trips. Geology isn't a discipline you simply pick up over the course of a few excursions into the desert, they grumbled among themselves; it's a lifelong vocation that takes a lifetime of study.

The scientists were careful not to share these misgivings with the astronauts, of course, but Aldrin, no fool, was well aware of what was going on. Determined to prove the turf-protecting scientists wrong, he fell to his geology studies with an almost inexhaustible zeal, becoming a lay expert in the field within months. Later that summer, when he and Armstrong found themselves in the Sea of

Tranquillity—a field site the boys in the lab coats had never harbored any hope of visiting and never would—he figured he was carrying the torch for every scientist-astronaut who would ever follow him there. Almost immediately, he appeared to have dropped it.

Armstrong and Aldrin's most important job after they climbed down the LEM's ladder was to collect and bag a few quick samples of rocks in case an emergency required them to terminate their lunar stay early. This so-called contingency sample was the highest-priority exercise in the early phase of the moonwalk, and the astronauts were expected to get to it straightaway. Aldrin, who climbed down to the surface only minutes after Armstrong took his first historic step, needed almost no time at all to get his bearings before setting to work. Poking about the landing site in the immediate vicinity of the ship, he noticed a small, vaguely purplish rock, freckled with what looked like bits of crystal. Buzz had seen a lot of samples just like this one back on Earth, and he did not hesitate to announce his find now.

"Hey, Neil," he called out. "Didn't I say we might see some purple rocks?"

"Find one?" Armstrong asked.

"Yep. Very small. Sparkly. I would make a first guess at some sort of biotite, but we'll leave that to further analysis."

To most people watching at home, Aldrin's observations were innocuous ones. But to the geologists in Mission Control, they were an absolute howler. Biotite! On the moon? Had these silly flyboys learned *nothing?* Biotite contains hydroxyl and hydroxyl forms only in the presence of water, a substance that was utterly absent on the rock-and-powder moon. You might find a lot of interesting samples in a place as geologically rich as the Sea of Tranquillity, but you had about as much chance of finding biotite as you did of finding a jackrabbit.

Throughout the scientific back rooms at NASA, knowing smiles and rolled eyes were exchanged all around, but among the astronauts, Buzz's blunder was seen as no blunder at all. If there was one rule the scientists had stressed to the astronauts from the very first day of their training, it was that the most important job of a field geologist is to observe and describe. Describing isn't analyzing; describing isn't identifying. Describing is merely telling the scientists back home exactly what you're seeing, and doing so in the clearest terms possible. If you found a rock that looked like a tangerine, you didn't waste your breath describing its color, shape, or nubby-rough surface. You said it looked like a tangerine and moved on to the next sample. It was the astronaut's job to explain what a rock resembled; it was the geologist's job to explain what it was.

Buzz had followed that rule, and had done so at the expense of his very credibility. From that moment on, every other lunar astronaut made it a point to mind his words, lest he be the next one to give the boys in the back rooms reason to laugh. Now, a little over two days after arriving on the surface of the moon, the astronauts of *Apollo 15* had their own remarkable sample to describe and their own phrasing to consider.

At ten o'clock on Sunday morning, Scott and Irwin found themselves standing three miles away from their landing site at the foot of the towering Hadley Delta Mountain. Three miles was a lot of ground to cover anywhere, never mind on the moon, and on *Apollos 11, 12,* and *14,* trying to trudge that far from the LEM would have been out of the question. On *Apollo 15,* however, the crew had brought a little help along: a nine-foot-long, four-foot-wide, open-chassis rover in which they could scoot across the surface at speeds approaching nine miles per hour. Unstowing the collapsible car early yesterday, they had hopped aboard and spent the better part of the last two days tooling here and there, looking for likely sites to stop and prospect. Just a few minutes ago they had parked on the lip of Spur crater and climbed down from the rover to pick through the rocks that littered its rim. Almost immediately something caught Irwin's eye.

Like all of the other *Apollo* crews who had come out this way before, one of the first things that struck the *Apollo 15* astronauts when they arrived at the moon was its almost total lack of color. Here and there the prevailing gray might give way to white and the white to a pallid tan or even a very faint—and very rare—purplish. But if there was any real color at all—any of the voluptuous reds and yellows and blues of Earth—it was only what was imported aboard the machines that brought the crewmen here. Yesterday, however, as Irwin ran his eyes over a slope near the lander, he could swear he saw something green flashing back at him. Irwin pointed this out to Scott, who saw the colorful shimmer, too. But when the astronauts inspected the ground more closely, the reflectivity of the soil—its albedo—faded to the familiar gray. Now, one day later and three miles away, Irwin was being teased by green again. He toed the soil beneath his feet and shifted it around in the sun. The color didn't change. Warily, he called out to Scott.

"I think we kicked up some more green material here, Dave," he said.

"Sure it isn't that light gray albedo stuff?" Scott asked.

Irwin squinted through his helmet at the ground in front of him. "No, it looks green."

"I think it might be the contrast."

"No, no. I see white, I see a light green, and I see a brown."

Scott bounced over to the spot where his junior pilot was standing, scraped at the soil with his rock rake, and picked up a single stone in the teeth of the tool. Bringing it close to his visor, he peered at it and laughed.

"I've got to admit," he conceded, "it really looks green to me, too."

Scott rubbed at the surface of the stone. While the soil flaked away, the color didn't. He turned it this way and that and the sample continued to glint at him. What was doing the glinting, it appeared, was a freckling of tiny, shiny, glassy beads embedded in the matrix of the material. Scott was impressed despite himself.

If the commander was reading his geology right—and he was pretty sure he was—these beads were volcanic beads, and not just from any volcano, but from the explosively dramatic variety known as a fire fountain. In volcanically active places like Hawaii, percolating magma not far below the surface could sometimes accumulate in a single, high-pressure reservoir, causing the crust above it to spring a pinpoint leak. When it did, the magma below would hiss out in a fine, aerosolized mist, discharging in all directions and settling on the surrounding land in a hellish sprinkling. When the droplets cooled, they hardened into small glass beads, and if the beads contained the right elements—principally magnesium—they turned green. Now, Scott was holding a specimen in his hand suggesting that the same exotic process had taken place on the moon.

Scott stashed a few of the glittery rocks in a code-numbered sample bag, taking care to read off the bag's numeral for the Mission Control scientists who would need the geologic dog tag to help them identify the specimen when the crew got home. The two astronauts then turned to a different piece of business. Immediately after arriving at the site, Irwin had spotted another curious-looking rock—a pale one this time—about the size of his ungloved fist. The rock was conspicuously perched on a small pinnacle of dirt a few steps away—so conspicuously, in fact, that to a romantic the pinnacle might even look more like a pedestal, a means of deliberately distinguishing this solitary sample from all the other ones scattered around it. Irwin had pointed out the rock to Scott, and Scott had promised to get to it later. Now, that time had come.

"Okay," Scott said summarily, "let's go get that unusual one."

Scott and Irwin bounded a step or two over to the rock and Scott bent toward it. He inclined his head this way and that, checking to see how the sun played off the stone, and immediately, a pearly flash caught his eye.

"There's a little white corner to the thing," he mused.

Scott lifted the rock to his visor, brushed away a bit of dust, and as he did, both men reacted with a start. The sample, they could see, wasn't just partly white, but almost entirely white. More important, it was a *sparkling* white. Winking up at them through the billions of years of lunar dirt was a constellation of diamond-like crystals, far brighter than the pedestrian glass that had studded the last sample. Instinctively, Scott started to blow more of the dust away and then caught himself when his wind was stopped cold against the inside of his visor. Instead, he simply thumbed the dirt off the rock; the crystals, now reflecting more sun, shone brighter still.

"Oh, man!" Irwin said.

"Oh, boy," Scott answered.

"I got . . ."

"Look at that!"

"Look at that glint!"

Scott turned the rock slowly in his hands and, for a moment, he could manage only one other sound: "Ahhh."

This, he knew instantly, was it. Ancient anorthosite, made up almost completely of plagioclase—a scrap of the original crust of the original moon if ever there was one. Over the course of 4.5 billion years of lunar history, meteoric violence had regularly shaken the moon, gouging out continent-sized seas all over its surface—yet this bit of rubble had remained unchanged. Over the same period, nearby Earth hardened and hydrated and evolved into the planetary petri dish that supported millions of forms of life—and a quarter of a million miles away, this bit of crust remained unchanged. Over that same time, the very age of the very universe increased by half—going from ten billion years old to nearly fifteen billion years old—and while young stars switched on and old stars switched off and nebulae and novas burst into life or exploded out of existence, this bit of crust remained unchanged. This was a shred of the primal moon, cooked up in the days of the primal solar system, and Dave Scott, an Earthly animal who had been nothing but elements himself a preposterously brief thirty-nine years ago, was holding it in his hands— and, much more important, *understanding* what he was holding.

"Guess what we just found?" Scott said to everyone and to no one at all. "Guess what we just found?" In the background, Houston could hear Irwin laughing happily. "I think we found what we came for," Scott said.

"Crystalline rock, huh?" Irwin asked, for Houston's benefit.

"Yes, sir," Scott answered. "You better believe it."

"Yes, sir," came the voice of Joe Allen, the astronaut in Mission Control handling the air-to-ground channel.

The precise significance of the rock neither Scott nor Irwin would say, not with the lesson of Buzz and his biotite still fresh in their minds. But none of the hundreds of people in Houston needed to be told what crystalline rock meant.

"Look at the plagioclase in there," Scott muttered to Irwin. "Almost all plage."

"That really is a beauty," Irwin said.

"Bag it up," Allen called from his seat in Mission Control as he watched the two astronauts on the movie screen–sized video monitor in the giant auditorium. On the screen, the remarkable rock was little more than a pixellated smudge.

Scott pulled out a fresh sample bag, started to place the rock inside, and turned to Irwin. "Let me get some of that clod there," he said, pointing to the mound of dirt on which the stone was found. As soon as he said that, though, he reversed himself. "No, let's don't mix them." He took a quick look at the bag's serial number. "Make this bag, 196, a special bag."

"Yes, sir," Allen answered from Earth.

Scott and Irwin dropped the remarkable rock into bag 196, cinched the now-precious gunnysack up, and kept it close to them throughout the remainder of the day. Even before the men of *Apollo 15* returned to Earth, reporters covering their mission had already dubbed their treasured find the Genesis Rock.

January 1973

Nobody talked much about what would happen if there was a fire in the moon rock building at NASA's Manned Spacecraft Center in Houston. Talking about a fire meant you were thinking about a fire, and thinking about it was something you didn't especially want to do. There were a lot of buildings NASA could lose and easily replace on the campus of its giant Houston facility, but the moon rock building wasn't one of them.

The moon rock building was originally and less colloquially known as the Lunar Receiving Laboratory, or the LRL, and its name did a good job of describing what it did. Well before the return of *Apollo 11* with its first, precious 47.7-pound cache of moon rocks, NASA knew it would need a place to store and study the rocks—a place that could keep the ancient samples as clean and pristine as they had remained in the endless eons they lay on the surface of the moon. Since no such lab existed on the planet, NASA built one, equipping its rock facility with airtight sample rooms, airtight meeting rooms, and negative pressure common areas that would allow clean air to flow out without allowing contaminated air to flow in. The locks on the vaults that protected the samples would be impervious to even the nimblest criminal fingers; the foundations on which the building was built would withstand even the most robust hurricane-force winds; the building itself would be all but entirely fireproof.

NASA officials were happy with their state-of-the-art lab—most of the time. The problem was, it was always possible to conjecture a fire or other disaster the likes of which no one had ever conceived, and more than one agency administrator lay awake nights fretting about just such a calamity. In order to provide the agency a little accident insurance—and NASA chiefs a little peace of mind—a small sampling of the lunar specimens was shipped out to a fortified military base, a fully secure place whose exact address was never disclosed to the public. The rest of the rocks would remain forever at the moon lab—and it was here that they would slowly begin to give up their secrets.

The studies conducted in the Houston facility generally began well before there was anything there to study at all—when the moon rocks were still on the surface of the moon. Geologists chosen to analyze the samples a lunar crew brought home would spend as much of the flight as they could camped out in a back room in Mission Control. As the astronauts in the field went about the task of collecting their samples, the scientists at the space center would keep a constant eye on them via the video cameras the crew took with them to the moon. Each time an astronaut came upon a sample worth collecting, the geologists would note precisely where it was found and in precisely what condition. When the rock was picked up and stowed in its coded bag, the geologists would record the bag's number on an index card—one card per rock—and set it aside. During the *Apollo 11* mission, with its comparatively modest load of samples, the pile of cards was a short one. By the last lunar landing—

Apollo 17 in December 1972, with its 243.1 pounds—the stack grew dramatically.

While most of the stones the moonwalkers discovered were collected and catalogued this way, not all were. Now and again, the crewmen would stumble across a rock that looked so promising it deserved special treatment. Perhaps it was covered with a dusting of unusual volcanic soil; perhaps it glittered an improbable shade of green; perhaps it was a bit of primordial crust that had lain undisturbed for several billion years before the astronauts stumbled across it. Whatever it was that set the rock apart, the idea of simply dropping it in a gunnysack and tossing it into the lunar module did not seem like a good one. Once the door to the LEM was closed and its cockpit was filled with air, the atmosphere inside the ship—filled with everything from bacteria to food debris to salt to spores to aerosolized urine—would seep through the bag and contaminate whatever was inside. In general, geologists were prepared for this and taught themselves to recognize and ignore out-of-place organic chemistry on their inorganic specimens. For their most delicate samples, however, they didn't want to take any chances.

Stowed aboard the LEM was a small stack of metal boxes measuring thirty inches across and just twelve inches high. When the astronauts found a specimen that warranted special care, they would carry one of these boxes out onto the lunar surface and open it up. If the box contained any stray atmosphere, it would burp it out into the lunar void, leaving behind the same near-perfect vacuum found everywhere else on the moon. The bag containing the sample would then be placed inside the box and the lid would be shut. When the box was carried back inside the LEM and the spacecraft was pressurized, the five pounds per square inch of cabin atmosphere would bear down on the lid of the chest, squeezing it tightly shut and causing its lining—made of an alloy of soft, ductile metals—to form an airtight seal. Back on Earth, where sea level pressure is an even more oppressive 14.7 pounds per square inch, the seal would become stronger still. Only when the box had been transferred to a larger, enclosed case in the moon rock building—a case filled with inert nitrogen—would the lid at last be pried open and the unlovely treasure revealed.

Typically, when a sample of this kind was being unveiled, it wouldn't be unveiled alone. The box would be opened by a single technician who would manipulate the specimen by placing his hands into a pair of heavy rubber gloves built into the side of the nitrogen case. A glass window on top of the case afforded him a view of what he was doing, and a far larger window built into one wall of the sealed room in which he was working afforded a similar view to the knot of geologists who would gather in the corridor outside. When the sample was first extracted, it would sometimes be covered with a layer of lunar dust, preventing the geologists from getting a good look at it. The technician in the room would then obligingly clean the specimen with a few gentle breaths from a nitrogen gun; as the billions of years of accumulated dirt swirled away, the watching scientists would get their first view of the crystals or glass or spanglings of color that had first caught the moonwalkers' eyes.

Painstakingly, over the course of several years, one after another of the 838.2 pounds of samples brought back by *Apollos 11, 12, 14, 15, 16,* and *17* were lovingly displayed and inspected this way. Then, less lovingly but no less carefully, they were sliced into more than 80,000 samples and biopsied down to their last constituent grains. The story those grains told was an extraordinary one.

For all of the assumptions the scientists and pilots of NASA made about the moon before the *Apollo* missions were launched, a lot of what they believed was based on *just* assumptions. One of the most enduring questions that had to be resolved concerned not lunar rocks, but lunar soil. If meteors indeed played a major role in shaping the surface of the moon, it wouldn't only be *big* meteors that did the job. Far more numerous than the giant space rocks able to dig out an entire sea would be the smaller, dust-sized bits of debris that bombard the lunar surface in a sort of constant cosmic rain. Over the course of billions of years, this persistent sand-blasting would work the soil over pretty completely, reducing it to a fine-sifted powder that was the stuff of geology ground small. So numerous were the collisions of these micrometeorites and for so long did they occur, that it was possible that the upper lunar crust was covered not merely by a thin skin of powder, but by a vast ocean of it that could, in theory, swallow up any machine or man foolish enough to dip so much as a toe in it.

Before the *Ranger* missions flew, Vienna-born astronomer Thomas Gold of Cornell University was the most conspicuously vocal scholar of the lunar dust theory. Gold had counted craters and estimated impact forces, and had come to the grim conclusion that the moon's surface was smothered in a blanket of dust that might reach down thousands of feet. The first crew of astronauts who landed on the moon, Gold warned darkly, would in all likelihood be the last, sinking immediately out of sight and remaining forever entombed in the sarcophagus of their lunar module.

Gold's theory suffered a blow when the three successful *Ranger* spacecraft returned close-up images of a lunar terrain that was littered with rocks and boulders—not the kind of heavy debris the cameras were likely to have spotted on a world whose fluffy surface had so much give. Advocates of the dust theory were undeterred by these images, arguing that if some moon rocks were porous and light, and if the soil they lay on top of was compressed in spots, there was no reason the occasional bit of debris wouldn't remain on the surface.

Things got worse for the Gold dust theory when the five unmanned *Surveyors*—which surely weighed more than a piece of rubble—landed safely on the moon in 1966, 1967, and 1968, and showed no sign of sinking. Even then, however, Gold's followers remained unpersuaded. With no absolute proof that the dust blanket was not miles deep, the spacecraft's creators had constructed it of the lightest materials possible and given it a broad, splayed-legged shape that would allow it to support itself on a thick layer of lunar dust much the way a swamp bug supports itself on water. If the *Surveyors* failed to sink out of sight, it might have said less about the state of the moon than about the ingenious engineers who built the ships.

It wasn't until July 1969 that Neil Armstrong finally settled Gold's hash when he put his *Apollo 11* lunar module safely down on the Sea of Tranquillity and found that the footing below both his ship and his boots was as firm as it was on Earth. The dust he and other moonwalkers brought back to the LRL proved to be made up of an enormous variety of pulverized rocks, suggesting that the surface of the world had indeed been pulverized by just the kind of microcollisions Gold and others had hypothesized. The force of those collisions, however, had simply been less than what Gold had anticipated.

Deep dust layer or not, the discovery that the moon had indeed been subjected to a rain from space sufficient to grind its surface to powder was monumentally important, settling not just the question of how the lunar soil was formed, but—in all probability—how lunar craters were formed. After several centuries of debate, there was still no absolute agreement on the process of crater formation, with the majority of scientists concluding that meteors were responsible. The new soil findings—illustrating just how ravaged the lunar surface was—gave this theory a considerable boost. Nonetheless, there was still a vocal minority that insisted that, when it came to midsized craters at least, it was not meteors that were responsible, but volcanoes. The dissenters' arguments were not totally without merit.

Earth has its own fair share of craters, virtually all found at the top of volcanic mountains or on the surface of sprawling volcanic fields. The floors of most of these craters are usually comparatively smooth, with the exception of a solitary cone made of cinder and ash that accumulates in the center. Even relatively weak telescopes had long revealed similar structures at the center of lunar craters, and close-up photos returned by manned and unmanned spacecraft had confirmed that discovery. More tellingly, on Earth, large craters often do not form individually, but rather are surrounded by clusters or chains of smaller craters, caused by underground magma percolating to the surface in several spots at once. On the moon, almost all of the largest craters appeared to be surrounded by secondary craters, too, and the closer astronomers looked—first with telescopes, then with spacecraft—the more numerous the smaller pits became. Finally, and perhaps most persuasively, large lunar craters were frequently found to be circled by concentric, rippling rings, as well as by long, sunburst-like rays that extended from the central pit in all directions. Similar surface scars are often found around volcanic craters on

Earth, caused by moving lava undulating away from a volcano or fast-moving lava splattering explosively out of it.

If lunar craters posed unresolved questions before the *Apollo* astronauts brought home their samples, the moon's color and shine posed an even greater one. Astronomers studying the face of the moon had always been puzzled by the curious way it reflects sunlight. Aim a light source at any sphere—from a planet to a ball to a soap bubble—and the illumination will not bounce back evenly. Rather, the center of the sphere will reflect the most brightly, while the edges, which curve away from the object emitting the light, will fall partly into shadow. The moon, however, doesn't do that. When the great white platter of a full moon is hanging in the sky, it beams back sunlight almost the way a real platter would—with its rim and edges reflecting precisely as brightly as its center does. Rationally, astronomers knew that when they were looking at the moon they were looking at a sphere; from the evidence of their eyes, however, they might as well have been seeing a two-dimensional decal slapped against the ceiling of the sky.

Just as perplexing as lunar light was lunar gravity. On Earth, gravity is an almost perfectly consistent thing, with the great mass of the planet pulling equally hard in all spots at all times. Place a hundred-pound block on a scale in, say, Mexico City, and you can be certain it will register the same hundred pounds if you later weigh it in Ottawa, Bangkok, or Berlin. Not so on the moon. When the *Lunar Orbiter* probes were flying their missions, NASA controllers often noticed that as the ships moved through their circular flight paths they would occasionally dip or skid a bit, almost as if the lunar gravity had suddenly tugged on them a little bit harder and then let go with a silent twang. During the flight of *Apollo 16* in April 1972, these gravity flutters proved to be so strong that a tiny subsatellite the crew released into orbit to study cosmic radiation soon began to wobble and fail, crashing onto the surface just two weeks after its launch.

NASA, which knew a thing or two about gravity, realized that such anomalies had to be caused by mysterious concentrations of mass beneath the surface of the moon, and appropriately enough began calling the patches of high-density matter mascons. But naming a thing is not the same as explaining it, and whatever the scientists chose to call the formations, they couldn't begin to account for them.

It was only when geologists began chipping answers out of the rock samples the *Apollo* astronauts had fetched home that most of these mysteries at last got solved. *Apollo 15*'s Genesis Rock, while not among the first samples brought back from the moon, was among the most precious, and as scientists studied its twinkly innards, it yielded just the primal secrets they hoped it would.

One of the first things NASA looked for in analyzing the Genesis specimen was the presence of strontium, an alkaline metal thought to have been present in certain known concentrations during the early, gestational stages of the solar system. If the rock contained strontium in the same ratios, the scientists would know that it was indeed an artifact of the original moon, accreting directly out of the hot, cosmic goo that made up all the local worlds. Chiseling off a tiny fragment of the rock and melting it down, they immediately detected the chemical signature indicating that strontium was present. Measuring how *much* strontium was being given off, they found it matched the levels in the early solar system perfectly. The Earth and the other planets, astronomers had always known, were about 4.5 billion years old; now they knew the moon was, too.

Once the geologists had established the age of the Genesis sample, the next step was to compare it to the age of the rocks the *Apollo* astronauts had brought back from the moon's seas to determine if they were indeed created later in the lunar life span. While the strontium method was a reliable way to conduct this further dating, it wasn't the only way. Another, more precise method—particularly when you're dating lots of different samples that could be lots of different ages—is to look at its isotopes.

Assuming the seas were indeed created by mammoth meteor strikes, the energy released by the impact of the projectiles would not only have vaporized the incoming rock itself, but also liquefied a thin layer of ground beneath it. When land is shock-melted this way, it undergoes a curious atomic change. Radioactive isotopes in rocks and soil tend to decay at a certain, fixed rate. If the land gets hot enough fast enough, however, the isotopes cook down into an earlier, more primal state, causing the decay to start all over again. In effect, the isotopic clocks deep inside the stones become reset, causing the fused, melted land within the freshly dug pit to become atomically younger than the crust around it. If you find a shiny layer on the floor of a sea or crater, you can be pretty sure the pit was dug by an incoming meteor; if you read the isotopes frozen in the soil, you can calculate precisely when that meteor hit.

As *Apollo* astronauts picked about at the edges of seas, they found a wealth of shock-melted rocks beneath the lava that otherwise covered the basins. And as geologists trained their radioactive sniffers on the samples, a variety of ages revealed themselves. Samples from the Imbrium basin, adjacent to the crustal zone where the Genesis Rock was collected, appeared to be 3.84 billion years old, meaning that the crust from which Genesis was blasted had existed undisturbed for at least the first 660 million years of the moon's existence before the sea was created and the rock was deposited on its banks. The distant Sea of Nectar, in the moon's southern hemisphere, appeared to be older still, having been dug from the crust 3.92 billion years ago; the Sea of Serenity, hard up against Imbrium, was 3.87 billion years old; the Sea of Tranquillity was 3.84 billion; both the Ocean of Storms and the Inner Sea were a comparatively young 3.16 billion. After that, little other evidence of major impacts turned up until the Copernicus crater was blasted into existence 800 million years ago, and the Tycho crater was created a scant 110 million years back. For the first billion or so years of the moon's history it thus appeared that the cosmos had administered it a savage beating, pummeling the new world with all matter of loose geological change that had yet to accrete into worlds. After that, things quieted down considerably, allowing the injuries on the lunar surface to heal and the globe as a whole to remain relatively peaceful.

With the evidence of the isotopic clocks confirming that both the moon's seas and soil were formed by meteors, the scientific majority arguing that all lunar craters were created the same way closed in for the kill. Studies of nuclear weapon test sites on Earth revealed that the main crater dug out by the blast was often surrounded by a series of smaller craters, gouged from the surface by shrapnel and other debris that the exploding bomb unleashed. A similar scatter pattern on the moon would likely be a result of similar secondary impacts caused by meteor debris rather than by secondary lava vents caused by volcanoes. The rings that surrounded the big craters could be explained by meteors as well, and were probably the remains of topographic rippling that occurred when the projectile that dug out the crater caused the ground around the impact site to liquefy temporarily; the rays that projected from the center of the craters were similarly the scars of backsplashing surface material blasting away from the point of the collision. Even the apparent cinder cones at the centers of the pits were, the geologists concluded, not cinder cones at all, but rather rebounded crust that was driven up by the meteoric impact and then became frozen in place. With these arguments—and the evidence to back them up—the volcano advocates at last threw in the towel, accepting that apart from the occasional fire fountain, volcanism had simply not played much of a role in the formation of the moon.

When that debate was at last settled, the geologists were free to move on to other, less contentious matters. One thing that engaged their imagination almost immediately was the chemical composition of the shock-melted rock layer the astronauts scraped up from beneath the lava bed. This glassy material, they found, turned out to be extremely rich in iron and extremely poor in aluminum. Samples from the surface of the nearby highlands seemed, curiously, to be made up of just the opposite stuff—lots of aluminum and little iron. Superficially, this didn't make sense, since a moon that jelled from what was likely a uniform glob of hot matter should be a lot more homogenous than that. The geologists, however, had long suspected that the early moon was only homogenous for so long. As the molten mass that forms any rocky body cools, it tends to stratify, with heavy metals precipitating down and lighter ones staying buoyantly on top. Over time, the cooling, hardening moon would have become highly layered, and when a big projectile like a meteor hit, the heavy geological viscera from far below would be dug out and scattered across the surface of the sea. The more of the deep, iron-rich stuff the *Apollo* astronauts found in a basin, the more violent they would know the sea-forming collision had been.

This same multi-layered lunar anatomy helped explain another of the moon's mysteries: the location and distribution of its seas. Ever since the Soviet Union's *Lunik 3* beamed back the first images of the far side of the moon in 1959, cartographers had been wondering why all of the large lunar seas were on the side of the globe facing Earth. Certainly, the far side had its share of craters—some of them monumentally big—but they were nothing *more* than craters, while the big, hemisphere-dominating plains were entirely absent. The reason for the disparity, the geologists now figured, almost certainly involved gravity.

Though the gravitational attraction of Earth tugged on the entire body of the moon at all times, it would be likely to tug a little harder on its dense inner layers than on its light, outer ones. This would cause the lunar innards to shift slightly Earthward inside their crustal shell, and that, in turn, would cause the crust to be slightly thinner on the side of the moon facing Earth than on the opposing side. A meteor that hit the moon's visible hemisphere thus had a greater chance of punching through the crust and getting a good, sea-creating lava flow going than a meteor that hit the *in*visible hemisphere.

Once the geologists understood the formation and location of the seas, they, in turn, found it easier to understand the mysterious underground mascons that so destabilized satellites orbiting the moon. It had not escaped the notice of NASA's orbital mappers that the most dramatic mascon tugs generally occurred over the biggest seas. As lava accumulated at the site of an impact, they now figured, it would cool, contract, and grow denser; as it grew denser, it would sink. Even a slight increase in the density of the surface commensurately increases gravity. An exponentially greater increase leads to an exponentially greater pull. After oozing so much lava over so many millions of years, the moon's seas had grown surpassingly dense, and three or four billion years later, when tiny machines from Earth started flying overhead, the vast lunar plains pulled on them hard indeed.

Just as important to the *Apollo* scientists as ancient geologic violence on the moon was the possibility of current geologic violence—specifically moonquakes. No one knew how seismically active the moon was, but in order to learn, the geologists had equipped all of the *Apollo* landing crews with portable, radioactively powered seismographs that could be left behind on the moon to operate unattended for years. Between 1969 and 1972 six of these sensitive instruments were deployed on the lunar surface, forming a pair of seismic triangles that covered much of the northern hemisphere. The positioning of the instruments allowed geologists on Earth not only to detect moonquakes as they occurred, but to pinpoint exactly where in the moon's interior anatomy the disturbance originated.

To be sure, listening for quakes on a world 250,000 miles away was more complicated than simply switching on the seismographs and waiting for their needles to jump. In order for the instruments to work properly, scientists first had to calibrate them, making sure they weren't tuned so low that they missed the moon's softer geologic whispers or so high that they mistook a mere tectonic twitch for a globe-jolting spasm. The problem was, there was no reliable way to adjust the machines unless the scientists could somehow predict exactly when a moonquake would take place and exactly what its intensity would be, then listen to how the seismographs picked it up and set their volume levels accordingly. While the moon itself could not provide this kind of seismic predictability, NASA could.

After a lunar crew lifted off from the moon and docked with the orbiting command module, the LEM became immediately expendable. Once the astronauts cut the ship loose, all it would take was a small push from the lander's nearly spent thrusters to destabilize it and send it tumbling down to the surface. An impact by a projectile that big, moving that fast, would certainly wake up the seismographs, and since scientists knew the spacecraft's weight down to the last ounce and its speed down to the last foot per second, they could easily calculate how high the needles ought to jump. Watching how much they did jump and then tweaking the machines as needed, they could prepare their instruments to spot any authentic tremors that might occur later.

As it turned out, not many did. In the eight years the *Apollo* seismographs functioned, they picked up almost no quake activity at all on the old, cold moon. Once in a great while an incoming meteor would sting the skin of the world and, like a horse twitching off a fly, it would shudder a little in response. Just as infrequently, the moon would stir a bit from within, emit a tiny seismic burp, and then settle back into slumber. So small and infrequent were these moonquakes that even if all of them were combined and released at once on Earth, they would be almost impossible to detect. The moon, it appeared, may have been a hot, seething place once, but it had long since gone still and cadaverous.

This very quiescence, it turned out, was also at least partly responsible for one of the moon's final unsolved mysteries: its unusual reflectivity. As the *Apollo* astronauts closed in on the moon during their outbound journey, they could not help noticing the same uniform brightness that had so long mystified astronomers. Whether they looked straight into the center of the lunar bull's-eye or off toward its most remote horizon, the light of the sun bounced back at them with an unshadowed brilliance that never varied. It was only when the astronauts actually landed their LEM and began trudging across the surface that they discovered the reason.

In the dry, low-gravity environment of the moon, dust particles don't accumulate in the comparatively dense blankets they do on Earth. Instead, lunar powder kicked into the sky—generally as a result of meteorite impacts—tends to land softly and arrange itself haphazardly, accumulating in loose mounds like fallen leaves. Covering the surface from pole to pole and horizon to horizon, these lightly packed dust bits play funny tricks with sunlight. Even as the edges of the moon recede away from the brilliance of the sun, some of the dust particles, propped up at all manner of angles, point back toward it, reflecting the light and obscuring the shadows beneath them. Thus, rather than retreating into shadow, the lunar horizons remain brilliantly lit, causing the entire face of the moon to shine back into space like a single celestial headlamp.

Such fairy castle soil, of course, is a fragile thing. Wind, if there were wind, would easily blow it to nothing; rain, if there were rain, would reduce it to mud. The merest touch from a gloved hand or a booted toe—to say nothing of the landing pad of a lunar module— would flatten it to a dirt mat. The delicately stacked particles could thus be witnessed by astronauts and even photographed up close, but trying to move them intact from the spot where they lay, much less bring them back to Earth, would be forever out of the question.

The sleepy stillness that allowed the fairy castle soil to last for so long was, of course, a dramatic departure from the violence in which the moon was born. Though meteors tattooed the lunar surface for the first billion or so years of its existence, in all the eons since, the cosmos appeared to have pretty much forgotten about the moon, setting it aside as little more than a dead, paperweight world in a solar system filled with far more dynamic places. Cosmic corpse or not, however, the moon was undeniably there, and the only question left unanswered by all the elegant data being pulled from the rocks and read from the seismographs was *how* it came to be there. That had the scientists stumped.

One of the prevailing truths about the solar system's moons is that no matter how different they look, they are without exception relatively small. Jupiter, Saturn, Uranus, and Neptune, measuring between 30,000 miles and 88,000 miles in diameter, are circled by swarms of moons that, compared to the vast size of the planets themselves, are little more than orbiting fruit flies. Mars, one of the solar system's smallest planets, is attended by two surpassingly tiny moons, one of which is less than fourteen miles across, the other of which is less than nine.

Earth's satellite breaks this size rule—at least in relative terms. Measuring 2,155 miles in diameter, the moon is about the same size as the largest satellites that circle the largest planets; its own parent world, however, is barely 8,000 miles wide. Such a one-to-four size ratio between a moon and a planet is rare indeed, making the two bodies look less like a dominant world and a dependent satellite than a pair of strange sibling globes orbiting each other as cosmic co-equals.

Also curious is the inclination of the moon's orbit. Unlike most large, natural satellites, which inscribe a more or less horizontal path around the equatorial waistline of their planets, Earth's moon appears to have been knocked slightly cockeyed, crossing 16 degrees above the equator at some points and 16 degrees below it at others. The orientation of the two worlds' poles is also a bit skewed. Earth is famously tilted at a near-drunken angle of more than 23 degrees, a list that accounts for the planet's wildly changeable seasons. The moon, on the other hand, is almost perfectly upright, tilted at just 1.5 degrees.

The existence of such a large moon at such an odd angle around Earth was never easy to explain. But even before the *Apollo* missions flew, a few lunar scientists were pretty sure they had the answer. Earth's moon, they figured, was not formed side by side with the planet it circles, but was instead taken captive by it. Racing alone through the outer solar system, the prehistoric moon was once a rogue world utterly untethered to any planet. On one of its mad passes through the planetary neighborhood, however, it happened to fly too close to Jupiter, allowing the massive gravity of the star-like world to flick it in toward the sun. Hurtling ahead on this suicide trajectory, the moon would surely have been consumed by the solar fires, but on its way toward the solar system's center it happened to pass by Earth, whizzing by just a quarter of a million miles away. Earthly gravity immediately grabbed the moon, yanking it off its straight, breakneck course and pulling it into a large, arcing one. Unsteadily at first, the moon began circling the planet. After a few wobbly laps, the new satellite stabilized itself, settling into a smooth, if forever-crooked orbit.

The theory was a good one, but only if you were willing to overlook a few critical flaws. First of all, a moon that had been flung sunward by a planet the size of Jupiter would be moving at a speed that was literally meteoric. Trying to stop a body so fast and massive would be a little like trying to catch a passing freight train with a fishing rod: You might get lucky and snag the thing with your hook, but the instant you did, the rod would be ripped from your hands. Earth was a powerful body all right, and the gravity field it emitted was impressive, but intercepting a target like the speeding moon was well beyond its powers.

Even if the moon were somehow moving slowly enough to slip into Earth's orbit, there was something about the newly discovered chemistry of the *Apollo* lunar rocks that also raised scientists' eyebrows. The three most abundant elements detected on the moon were oxygen, silicon, and aluminum, in that order. These elements in precisely the same order are also the most abundant ones found in Earth's crust. A free-floating world that formed deep in the cold ether of the distant solar system would not be likely to be made of the same elemental stuff as a body like Earth, which was cooked up millions of miles farther in, over the open fires of the nearby sun. The inclination of the moon may have suggested that it was a relatively new arrival in the Earthly neighborhood, but its chemistry and size suggested otherwise.

If Earth didn't capture the moon, it was always possible it gave birth to it. Early in the solar system's history, Earth, like the moon, was a molten body with a consistency less like rock than putty. Get a planetary mass soft enough and ratchet up its spin velocity high enough, and it just might fling off a globule or two. In the deep freeze of deep space, it wouldn't take much for one of those globules to harden into a moon.

This lunar fission theory had a lot of appeal—not least because it avoided the nasty gravitational problems of the capture theory. The trouble was, the sharp angle of the moon's orbit—the very angle that gave the capture theory whatever credibility it had—made the fission theory less than credible. A molten Earth that was throwing off debris would likely throw it off from its fastest-spinning point, and that was its equator. Matter flung from the equatorial plane would likely stay in the equatorial plane, not hop a random 16 degrees above and below it.

The only realistic alternative to both of these theories was that the moon neither budded from Earth nor was lassoed by it, but simply condensed from primal stellar matter along *with* Earth. If the planets were nothing more than leftover matter that failed to condense into the body of the sun, there was no reason that the moons couldn't similarly be leftover matter that failed to condense into planets.

This accretion theory had perhaps the most adherents of all, but it, too, had fatal flaws. As with the fission theory, the nature of the lunar orbit posed a problem, since planets and moons that formed from the same spinning cloud of matter would likely have the same angle and inclination. And as with the fission theory, once again lunar chemistry was a puzzle. While the moon is rich in the same oxygen, silicon, and aluminum so common on Earth, it has little or no volatile, easy-to-boil materials like water, sodium, hydrogen, and helium. Earth, by contrast, is fairly drenched in volatiles. If the moon hardened from the same raw cloud that created the planet, it ought to have virtually all of the same ingredients.

It wasn't until another twelve years had passed that lunar astronomers came up with a fourth—and, they hoped, final—theory, one that combined the best of the other three and left none of the same troublesome questions unanswered. When the solar system as a whole was just accreting, so the new scenario went, a protoplanet about the size of Mars was moving about the sun in an orbit similar to Earth's. Similar orbits, of course, can be deadly orbits, since if the trajectory of one world should drift by only a degree or two, it could directly cross the path of the other, with disastrous results. About 4.5 billion years ago that is precisely what happened, as the smaller of the two planets lost its way and cannonballed blindly into the flank of the proto-Earth. The collision was a slightly off-center one, gouging a massive divot from the larger world, tilting it slightly off center, and increasing its spin rate to one rotation every twenty-four hours.

While the young Earth was only wounded and dizzied by the blow, the smaller, incoming world was utterly destroyed. Much of its heavy, metallic mantle was incorporated into Earth's own mantle. The rest of it was liquefied, vaporized, and ejected into space, where it began to orbit Earth at an altitude of about 238,000 miles. The pulverized swatch of Earth that had been dug out by the impact drifted up and joined it there. For a time, this cloud of orbiting debris was nothing *but* a cloud. Quickly, however, it contracted into a ring, and just as quickly the ring contracted into a moon.

This new impact theory had a lot to recommend it. For one thing, it explained the similar oxygen, silicon, and aluminum content of the two worlds. Since a good portion of the new moon was made up of material from Earth, it was only natural that both bodies would contain some of the same elements. For another thing, it explained the absence of volatiles like water, sodium, hydrogen, and helium on the moon. The collision between the impactor world and Earth generated intense heat, more than enough to cause any materials with a low boiling point to evaporate entirely. On Earth, which largely survived the blow, plenty of these materials would have remained unaffected. On the moon, which was made up only of superheated collision debris, they would be completely depleted. The violence of the impact also explained the moon's curiously inclined orbit. Lunar material that was blasted from Earth, after all, would fly out a lot more explosively than lunar material that was gently spun from it, and it would be no surprise if the primal moon wound up outside the plane of its parent world's equator, moving through an orbit that perpetually took it first above and then below the Earthly midline.

That was the answer the moon scientists of the mid-1980s came up with—and that was the answer that largely stuck. But the moon scientists of the early 1970s—the ones who had first cracked open the rock boxes that the *Apollo* crews brought home—had not yet reasoned things through that far. The NASA administrators of that earlier era were not, to be sure, happy with these unresolved questions. The agency had spent more than $30 billion to send twenty-two manned and unmanned missions to the moon, and for that kind of money they expected some answers. The scientists who were charged with the job of coaxing the secrets out of the lunar rocks, however, were not so troubled. Over the course of three and a half years, nearly half a ton of the moon had been pried loose and carried back to Earth, and the stories those stones and dirt could tell would far outnumber the ones they couldn't. For the moment at least, most of the *Apollo* geologists seemed blithely untroubled by their inability to solve one lingering lunar riddle. Indeed, they often liked to tell one another with winks and grins, it was entirely possible the moon didn't exist at all. If it did, the theories they did have would explain it.

Part II

Far

3

The Grand Tour

Pasadena, Calif., Late December 1971

William Pickering had suspected for some time that the people in Washington weren't going to let him go to Neptune. The press had warned him; the NASA brass had warned him; even some of the junior members of his Jet Propulsion Laboratory staff—people who had no real business cautioning their boss about such matters—had occasionally presumed to warn him. Congress, however, had not yet said a word, so Pickering did not give the matter much thought. When the news at last came from Washington that Neptune was indeed out of the question, it thus came as something of a jolt.

Actually, it wasn't just Neptune Pickering had been thinking about visiting. Neptune was part of the itinerary, certainly, but it was only the last part. Well before Pickering's spacecraft ever got as far as the eighth planet, he had also planned to send it to visit the fifth, sixth, and seventh—Jupiter, Saturn, and Uranus—paying a call on the whole archipelago of large worlds from the mid-solar system out to the very end. Regardless of where Pickering intended his ship to go, however, according to the word that had just come from the East, it now looked as if it wouldn't be going anywhere at all.

To be sure, it wasn't that Pickering had had much cause to complain about the treatment he'd gotten from Washington over the years. Oh, there had been that nasty bit of business with the *Ranger* program back in 1964, when all of JPL's experimental moonships kept failing or falling long before they ever got where they were supposed to go and committees on Capitol Hill started calling for someone's head, and Pickering had had to serve up poor Jim Burke's and poor Cliff Cummings's just to appease the baying lawmakers until *Ranger 7* could at last fly the mission it was supposed to fly and effectively save the lab's bacon. But that was a long time ago, and it seemed unseemly to be thinking about that after all the places Congress had let Pickering fly his ships since. There had been the five *Lunar Orbiter* spacecraft JPL launched successfully to the moon in 1966 and 1967. There had been the seven *Surveyor* spacecraft sent to make soft lunar landings between 1966 and 1968. There had been *Mariner 2* and *Mariner 5*, which successfully flew by Venus in 1962 and 1967; *Mariner*s 4, 6, 7, and 9, which reconnoitered Mars between 1964 and 1971; and *Pioneer*s 7, 8, and 9, which traveled to the sun in 1966, 1967, and 1968. There was even talk of a *Mariner 10* next year that would fly past Venus, pick up a gravitational boost from the big Earth-sized world, and then fling itself on to little moon-sized Mercury—a single spacecraft that would visit two separate planets using no more hardware and no more manpower than it would usually take to visit just one.

What pleased Pickering especially when he thought about all these missions—and what, by rights, ought to please Congress, too—was how cheaply he had flown them. It had cost about $24 billion over the past decade for NASA to mount its manned space program, with the costs climbing especially fast in the last three years when the agency began launching its *Apollo* lunar missions, each one setting the government back a cool $400 million. From the beginning Washington had made it clear that the unmanned space program could expect barely a fraction of that kind of lavish spending, repeatedly reminding labs like JPL that if they were going to survive, they were going to have to do so on the table scraps and pan drippings *Apollo* left behind. Pickering had always taken pains not to complain about such comparative starvation rations, even making it a matter of pride to bring his projects in as far under budget as he could get them. Congress, however, had never seemed quite as impressed with Pickering's parsimony as Pickering himself, and no matter how tight the JPL director managed to pull his lab's fiscal belt, Congress always tried to make him pull it tighter still. Today, Pickering learned, Washington had tightened things up but good.

Pickering learned the bad news about Neptune the way he learned most other things that happened on Capitol Hill: through a twenty-something girl named Victoria Melikan. To anyone who worked in the Pasadena space community where Pickering made his living, a name like Victoria Melikan didn't mean much. If you did your space business in Washington, however, you got to know Victoria early and you got to know her well.

It was Pickering himself who had first spotted Victoria Melikan almost ten years earlier, when she was a University of Michigan undergrad picking up extra money working weekends and afternoons at the General Motors headquarters just outside Detroit. As pocketmoney jobs went, Victoria's wasn't a bad one: Half her time was spent writing speeches for one of the company's executive vice presidents, the other half was spent working in the company's government sales department, helping to administer some of the lavish research and manufacturing deals GM had struck with Washington. Even by college girl standards, the income Victoria was pulling down wasn't much, but that could change dramatically: GM, she knew, rewarded its employees well, and if she could just stay aboard until she finished school and then join the payroll full-time, she might get a chance to share in some of the riches, too.

One day in 1961, shortly before graduation, Victoria learned that a William Pickering from one of the new space labs in California would be dropping by GM for a technology conference of some kind. Space work wasn't strictly part of Victoria's job, but making important visitors feel welcome was, and she was given the job of showing this Mr. Pickering—actually *Dr.* Pickering, she kept reminding herself—around the plant. Pickering himself didn't really care who escorted him during what was essentially a glad-handing tour, but when he met the impossibly self-possessed young woman who had been assigned to his care, he was impressed by her immediately.

He liked her calm, he liked her poise, he liked the way she shook his hand respectfully but not too gravely. By the end of his visit, he knew he wanted her for his lab.

"Miss Melikan," Pickering said as he was taking his leave, "what are you doing wasting your time with automobiles?"

"I like automobiles," she answered.

"Nothing wrong with them," Pickering said agreeably. "Me? I prefer spacecraft. The way I see it, anyone can build a machine that will take you down a road. But there aren't too many who can build ones that will take you to the stars." Pickering reached into his pocket and gave Victoria his card. "You call me if you ever get restless here."

A year later, Victoria graduated from college and was, as she had hoped, hired by General Motors full-time, working in a position not all that different from the one she had held when she was in college. Two years after that, in mid-1964, during the beginning of what looked like a promising climb up the GM trellis, she indeed, as Pickering had suggested, began feeling restless. Pulling Pickering's by now dog-eared card from the back of her desk drawer, she phoned his office. To her surprise, he remembered her; to her greater surprise, he agreed to see her.

"I might want to take you up on that offer you made me," she said with little preamble when she presented herself in Pickering's office.

"What do you make at GM?" Pickering asked.

Victoria told him.

"I can offer you half of that."

"GM gives me a bonus every year."

"No bonus."

"GM lets me have a new car every three thousand miles."

Pickering suppressed a laugh. "No car."

"GM gives me—" Victoria started to say, but Pickering raised a silencing hand.

"GM gives you what private industry can give you," he said. "JPL will never match it."

Victoria paused. "Let me give it a night's thought," she said.

That evening, she went home and, well before the night was over, told her father that she'd be leaving her much-treasured job at GM and going to work for a Dr. Pickering at the Jet Propulsion Laboratory in Pasadena. With her first modest paycheck she'd have to start saving up for an equally modest car.

Victoria's initial assignment at JPL was in the public affairs department, but Pickering quickly decided that given her experience with government contracts, where he really wanted her to go was Washington. All of the NASA facilities across the country made it a point to have at least one full-time employee on Capitol Hill to lobby for funding for the agency's current and future projects. None of the NASA representatives ever called what they were doing lobbying, of course, and in a place as coy about such things as Washington, nobody expected them to. What they were really doing, they said, was simply explaining their programs, clarifying their programs, buttonholing congressmen in corridors, offices, and committee rooms not so much to press them for their votes, but simply to keep them abreast of the imaginative things the engineers and flight planners in the NASA labs were up to every day. If what they were up to interested Congress enough to write the agency a bigger check this year than it did last, well, that was just fine, too.

In the first few years the NASA representatives were coming to Washington, they didn't have to work very hard at their job. For all the lawmakers' resistance to being lobbied, the fact was, most of them seemed to like space. They liked the big, sleek rockets that often got built in their very own states, providing jobs for their very own constituents; they liked the grainy pictures the rockets sent home as visual proof that the public works money had been well spent; they liked the buzz-cut astronauts who sometimes visited their districts or spoke at their high schools and always made it a point to shake the congressmen's hands and remind the crowd that if it weren't for the vision of elected officials there would be no space program at all. The NASA representatives came to Washington prepared to sell their programs, and on the whole, they found lawmakers who were enthusiastically willing to buy.

At least most of the time they were. But Victoria and the other lobbyists noticed something else about the congressmen. It wasn't evident in anything they said; it wasn't evident in how they voted. But it *was* evident in subtler things: in the menacing noises they'd make when a program slipped over budget; in the lusty way they'd piled on JPL after all of the disabled *Ranger*s died. It was even evident when a mission succeeded brilliantly.

In the spring of 1966, when *Surveyor 1* was preparing to make NASA's first controlled landing on the moon, setting itself down on a puff of rocket exhaust in the same hard-baked plains in which its *Ranger* brothers had so explosively crashed, national interest in the mission was keen. As with the *Ranger* flights, the press had been following the mile-by-mile progress of the ship from the moment it was launched; and as with the *Ranger* flights, on the day the spacecraft was scheduled to land, the auditoriums and media rooms of JPL were fairly overrun by visitors determined to be on hand for so historic a moment. Unlike the *Ranger* landings, however, which were observed mostly by reporters and JPL employees themselves, this one was attended by an entire House of Representatives subcommittee— the Subcommittee on Space Science and Applications.

A visit by so exalted a group was a first for JPL, and it was Victoria who was assigned the job of looking after the committee men while they were there, escorting them around the lab's giant campus and sitting with them in the Space Flight Operations Facility's glassed-in VIP area during *Surveyor*'s final approach to the moon. Attending the landing in the company of so many men who were in a position to help or hurt JPL would not be easy, and throughout the day Victoria would be concerning herself as much with the welfare of the spacecraft as with the mood of the lawmakers there to watch it fly. It was in the final hour before *Surveyor 1* touched down on the moon that all of the VIPs at last settled into their seats in the SFOF viewing gallery, and after they did Victoria found herself preoccupied with trying to read their moods, regularly glancing at them for almost any cue that would reveal almost anything about their states of mind. With each announcement the mission narrator made—singing out the spacecraft's dwindling altitude and trajectory figures through the public address system in the gallery—she made it a point to catch the eye of this or that congressman, offering each an all-is-well smile or an A-OK nod. She got little in return.

Finally, at the end of the hour, when *Surveyor 1* was just a few feet above the surface of the moon, the spacecraft went into a brief, planned hover and the calls from the mission narrator stopped. For long seconds, no one in the VIP gallery or on the SFOF floor spoke, as the ship throttled its engine down and attempted to lower itself into the ancient soil. Then, all at once, the public address system again broke the silence as word arrived from the moon that the maneuver, against all odds, was complete, and the ship was safe.

"*Surveyor* is down," the announcer called exultantly.

Immediately, the gallery broke into an ovation, as hands were shaken, backs were slapped, congratulations were exchanged. Victoria, casting her eyes back to her congressmen, saw with satisfaction that they were joining the general celebration, too. Suddenly, however, the mission narrator, who should have had nothing more to say to the deeply relieved crowd, called out again.

"*Surveyor* is up," he said flatly.

To the people in the gallery who knew something about the mission, this was not unexpected. A *Surveyor* ship was an extremely lightweight machine and even with its engine breaking its fall it would still hit the lunar surface hard enough to cause it to bounce once or twice. A mere bounce still counts as up for a spacecraft that should be down, and mission rules required the narrator to announce the ship's position, no matter what it might be. If this *Surveyor* performed the way the design specs said it should, it would take only a second or two for the energy of the landing to be fully spent and the spacecraft to settle back down in the soil. A second or so later, that indeed happened and the public address voice quickly made the announcement that *Surveyor* was once again safe on the surface of the moon.

In that same second, however, Victoria noticed something disturbing about her congressmen—something in the set of their mouths, in the slight nods they exchanged, in the way they leaned expectantly forward like a steeplechase crowd when the lead horse all at once pulls up lame. Something, ultimately, in all of those things taken together.

They'd shut us down, Victoria suddenly realized. *Give them any excuse and they'd shut us down.*

Five years later, in December 1971, when William Pickering was trying to build a marvelous space probe that would visit Jupiter, Saturn, Uranus, and Neptune all in a single mission, Congress began doing just that.

•   •   •

It wasn't Pickering who figured out how to fly a single spacecraft past four different planets, and he never pretended to take credit for it. Just who actually did work out the plan was always a matter of some dispute, but the granddaddy of the thing looked to many people at the Jet Propulsion Laboratory to be Mike Minovich. By any measure, Mike Minovich—a newcomer to JPL who first showed up at the lab in 1966—was too young to be an authentic granddaddy; indeed, he was too young even to be a fully credentialed JPL scientist. What he was, at least when he came through the door, was a graduate student, just one of an eager group of apprentice researchers on loan from UCLA or Caltech, putting in a semester or two at the lab in the hope of being asked to join full-time when his classes were done.

Or that's what everybody said Minovich was, anyway. The thing was, not a lot of people ever saw him to confirm this. Oh, they saw his work, all right; it was impossible not to see Minovich's work, stacked up the way it was in giant mounds of computer paper flowing from his desk to his floor and into the hallway outside his office. But they just never saw what he did to generate it all. Come in in the morning and Minovich had already punched out for the day. Leave in the evening and he had not yet punched back in. When Minovich did like to work was at night, generally *all* night, a twelve- or fourteen-hour stretch when the JPL campus was quiet, the distractions were few, and the lab's room-sized mainframe computer was not being used, giving a simple graduate student the only chance he might ever have to get his hands on the mammoth machine.

Minovich didn't talk to anyone else who worked the night shift about what he and the mainframe were trying to puzzle out, but if you stumbled across one of his hummocks of paper crowding the hallway it was immediately evident what he was up to. The pages were filled with trajectory calculations—arcs and angles and complicated thrust numbers that were pretty standard stuff for a JPL researcher. But unlike most trajectory profiles, which are designed to get a ship from one point, typically the Earth, to another, typically the moon or Mars or Venus, this one was aiming for all four of the giant planets in the outer solar system.

The idea of making such a one-ship, four-world trip had been kicking around JPL for months before Minovich started working out the arithmetic of the journey, ever since Jim Long and the other scientists in the future missions department had discovered that sometime in the late 1970s, Jupiter, Saturn, Uranus, and Neptune were going to arrange themselves in an orbital conga line that would not occur again for another 176 years. Start planning right away, and you could launch a spacecraft in 1977 to arrive at Jupiter by 1979 and reach the other three planets over the course of the following ten years. Dawdle too long and you wouldn't get your next chance to make the same 2.7-billion-mile trip until 2153.

Puddle-jumping from world to world when they were queued up this way would have advantages beyond mere navigational convenience. A tiny dust mote of a spacecraft approaching a massive body like a planet does not just sail blithely by without feeling some of its effects. Well before the ship makes its closest approach to the world, the long arm of the planet's gravity reaches out to grab it. If the advancing spacecraft is traveling slowly enough, the gravity grip cannot be broken and the little machine will simply crash into the unyielding surface of the world—or, in the case of gas giants like Jupiter, Saturn, Uranus, or Neptune, simply be swallowed up by the planet's huge swamp of an atmosphere.

If the spacecraft is moving at enough of a clip and gives the planet a wide enough berth, however, even the most powerful gravity can't hang on to it. Rather, the ship will simply arc partway around the rear side of the world, surf its gravity field back around to the front, and then continue on its way into space. This gravitational whipcracking has two effects on the spacecraft: It bends its trajectory this way or that, depending on the mass of the planet and the proximity of the approach, and it speeds the ship up. Plan your approach to a planet you hope to visit carefully enough and you could use its gravity both to point your ship to its next destination and to step on the gas and get it there faster.

To be sure, a Newtonian free ride like this is not *completely* free. Since energy can neither be created nor destroyed, but rather merely exchanged, a planet that speeds a spacecraft up must itself slow down a bit. A ship whipping around the far side of, say, Jupiter at 22,000 miles per hour would appear from behind the planet's eastern edge moving at a speed of close to 35,000 miles per hour. In exchange for the kick a planet this large adds to a ship's speed, the planet itself must lose a little velocity, decelerating by about a foot of motion every trillion years. Though the loss is real, it was one most astronomers figured the big planets could afford.

While describing the theory behind such slingshot trajectories was a simple matter, plotting the trajectories themselves was infinitely harder. It was this job Minovich the grad student assumed for himself, and as the semester wore on and the obscure apprentice slowly began revealing his work to the scientists who supervised him, it became clear that such a mission could indeed be flown. The real trick would be building a spacecraft that was able to fly it.

A ship being designed to travel to the deepest provinces of the outer solar system would face a lot of engineering challenges, but the biggest one would be power. Probes setting sail for a target in the inner solar system—Mercury, Venus, Mars, or the moon—were remarkably lightweight machines, in part because they didn't have to carry much energy-generating hardware along. Flying in such close proximity to the bright, white fires of the sun, the ships needed to do little more to power themselves up than unfurl a pair of solar panels and drink up all the juice they needed for however long they intended to fly. On those few occasions when they flew around the shadowed side of a planet or moon, a small on-board battery could pick up the slack temporarily.

In the outer solar system, things are different. A planet that sits twice as far from the sun as, say, Earth does, doesn't just receive half as much sunlight. Rather, owing to the three-dimensional geometry of light dispersal, it receives barely a quarter as much. Three times greater distance causes sunlight to be squared down by a factor of 9; ten times greater distance means a hundred times less light. The planet Neptune, which floats at the end of a gravitational tether nearly 2.8 billion miles from the sun, receives nine hundred times less light than Earth, which lies thirty times closer, making high noon at Neptune's equator about as bright as a moonlit Earthly night.

Flying through such a cosmic gloaming makes solar panels patently useless. The obvious alternative—an on-board battery system—is equally impractical, since carrying enough batteries to keep a ship alive through the twelve years it would take to fly to a target like Neptune would add uncountable pounds to the weight of the ship and uncountable dollars to its price tag. If the deep solar system spacecraft was going to fly, it would clearly need juice, but just how it was going to get it was, at the moment, a mystery.

Just as problematic as the spacecraft's power system was its communications system. The transmitters aboard the earliest *Mariner* ships to Mars and Venus—to say nothing of the still earlier *Ranger* ships to the moon—were simple affairs, sending out data at a mere eight bits per second, or about the same rate as a Morse code key. Even traveling at the speed of light, that trickle of information could take up to eleven minutes to reach Earth from Mars. Assembling a single picture from a data stream this slow and thin took more than eight hours and produced an image that was patchy at best. In the provinces of the outer solar system, where there were four large planets and potentially dozens of moons to photograph and where light-speed transmissions could take hours to reach home, such a feeble system was out of the question.

Even more troubling were the spacecraft cameras. Photographing a planet or moon in the deepest shadows of the deep solar system was a tricky business, made possible only if the shutter of the imaging system was kept open for up to a minute at a time. This was easy enough to do with a camera that was standing still. With a camera aboard a spacecraft moving at a clip of 35,000 miles per hour, however, any image that resulted from such a time exposure would be little more than a dark, cosmic smudge. The only way to prevent that from happening was to invent a picture-taking system that could somehow compensate for the motion of the spacecraft, allowing the ship to pivot as it passed a planet so that it could stay photographically fixed on its target without slowing down by as much as a single mile per hour.

Until these and other technical problems could be worked out, a grand tour of the outer solar system would remain little more than the febrile fantasies of a Mike Minovich or a Jim Long, men with lots of time on the mainframe computer, but no real responsibility for turning their paper cipherings into functioning hardware. William Pickering did not care much for febrile fantasies, preferring that his scientists either fly missions or not fly them, but not spend untold time and money merely contemplating flying them. In late 1968, two years after mission planners began speaking dreamily of the grand tour, Pickering called a meeting with Homer Joe Stewart, the head of the future missions department; Robert Parks, the head of the lab's planetary projects department; and Clarence Gates, a trajectory wise man who worked for Parks. When the three men were assembled in his office Pickering got directly to the point.

"What exactly do we have with this grand tour?" Pickering asked summarily. "Is it a real mission or isn't it?"

"Clarence has worked the numbers," Parks said, "and they suggest it is."

Pickering turned to Gates, who glanced down at a sheaf of papers in his lap. "The alignment is real," he said. "The planets should move into position around 1979."

"And stay that way for how long?"

"We'd have until about 1989 to get through the system."

"Do we know how to do that?" Pickering asked.

"The physics and navigation aren't a problem," Stewart said. "It's the hardware that we need to work on."

"And have we worked on it?"

"Not yet."

Pickering paused. "Then," he said, "I suggest you gentlemen get started."

Before the end of the year Pickering had contacted NASA headquarters and requested funds to help the lab figure out how to build a spacecraft capable of embarking on a twelve-year mission to the outer planets. Weeks later NASA responded, not only approving preliminary funding for the project, but also approving a brand-new name for it. The proposed flight to the outer solar system that had been previously known as the grand tour would from now on be known as the TOPS project, for Thermoelectric Outer Planet Spacecraft. Such acronymic aridness was not pretty, but for a federal agency like NASA, it was the surest possible sign that the project had just moved from a mere idea to a fully certified mission.

●　　●　　●

It took three years—until the fall of 1971—before the plans for the TOPS spacecraft were at last completed, and it was clear to anyone who saw them that if the thing ever flew it would be over Congress's dead body. The problem was not that the ship wasn't a good one; indeed, even to the untrained eye of the legislator, it was evident that the TOPS spacecraft was a dandy machine. The problem was what it would cost to build it, and the way the blueprints looked at the moment, it would cost plenty.

The knottiest problem the JPL engineers had been confronted with—the power problem—was the one they were able to solve most easily, and the solution they came up with was nuclear. Radioactive material gives off heat as it decays, and a lot of radioactive material gives off a lot of heat. The TOPS designers proposed to fuel their spacecraft with twenty-four pellets of plutonium 238 separated into clusters of eight and packed into three iridium spheres. The thermal energy radiated by all of this material could be channeled into a small, on-board generator that would then produce electricity, much the way turbines driven by nuclear-heated steam generate power on Earth. The spacecraft power system—which the designers called an

RTG, for radioisotope thermoelectric generator—would be an impressively compact machine, only seven feet long and roughly cylindrical in shape. What's more, it would be an efficient machine. As the plutonium in the generator decayed, it would lose only about 2 percent of its total energy per year. Barring an unexpected breakdown, this would give the portable power plant roughly half a century of life, more than enough for the mission the JPL planners envisioned.

Just as impressive as the RTG was the on-board computer the engineers designed for their ship. For a spacecraft that had to operate trouble-free for more than a decade, the traditional system of a prime data processor and a single backup processor would not do. Rather, the TOPS was equipped with no fewer than five redundant computers—three prime systems and two backups. Before a critical maneuver was attempted, the trio of main processors would consult one another and decide how and when it should best be executed. If one of the three disagreed, the other two main computers would investigate why. If it turned out that the dissenter had discovered a flaw in the plan, its proposed fix would be adopted. If the third computer was simply malfunctioning, it would be taken off-line and one of the two backups would be promoted. The JPL engineers ambitiously called their system the STAR computer, for Self-Test and Repair, and they were convinced the hardware would live up to that handle.

The communications system for the TOPS was souped up, too. The spacecraft was equipped with a twelve-foot parabolic antenna that was able to send information to Earth not in the trickle the *Mariners* could produce, but in a comparative electromagnetic gush. Aimed at deep-space receiving stations in Australia, Spain, and the Mojave Desert, the transmitter could beam back information at a rate of up to 115,200 bits per second. Such a rich data stream would be able to carry a wealth of planetary information, including readings of electromagnetic fields, chemical spectra, cosmic rays, plasma emissions, and, of course, crisp color images. Taking those pictures would still be something of a challenge, since the engineers in the flight dynamics department still had not figured out quite how to pivot the ship so that its cameras could stay stationary long enough to do the work they were intended to do. But given the quality of the spacecraft the designers were presenting them with, few people at JPL doubted that this last bit of business could be figured out.

Picking a precise trajectory for the TOPS was a less complicated job than all this engineering and design work. Working with Mike Minovich's preliminary calculations, the flight planners had determined that in order to reach the outer planets when they were in their maximum alignment, the spacecraft would have to leave Earth during a narrow forty-day launch window that ran from mid-August to mid-September 1977. Once the ship got off the ground, there were more than ten thousand subtly different flight paths it could follow, depending upon the precise photographic angle mission managers wanted it to have as it approached any one planet. So flexible were all of the available trajectory options that the analysts concluded it might even be possible to include a fifth planet, Pluto, in the overall mission profile. Since Pluto would not lie in the same clean line as the other four worlds, no one spacecraft would be able to fly the comparative corkscrew trajectory necessary to visit all five. But if NASA was willing to build two ships in the TOPS class, the flight planners could target one to fly a relatively slow route past Jupiter, Saturn, Uranus, and finally Neptune, while another would travel first to Jupiter and then bypass all the other worlds, riding the big planet's gravitational slingshot straight out to Pluto.

Naturally, the cost for such a pair of Cadillac spacecraft would be considerable. Near as JPL budget makers could calculate, the TOPS ships would probably set NASA back at least $750 million. In a space agency that routinely spent $100 million or more for a single expendable booster, $750 million was arguably not all that much—not when it would produce two spacecraft and more than twelve years of planetary research. But that was $750 million before any unexpected hardware problems; $750 million before any unexpected design setbacks; $750 million before inflation and labor disputes and contract snags ratcheted the cost higher still. That kind of $750 million could easily turn into a billion or more and everybody in the space community knew it.

It was for that reason that Pickering and the rest of JPL figured that the TOPS program was probably dead on arrival in the halls of Congress. Certainly, it was always possible in situations like this to scale the project back before it ever got to Washington. But engineering programs weren't education or environmental or social services programs—programs that lent themselves to compromise and half-a-loaf concessions. Engineering programs had hard, empirical, non-negotiable answers, and when Pickering's designers presented him with those answers in the form of the blueprints for the TOPS ship, his scientific instincts told him that these were the plans he ought to submit to Congress—even if his political instincts told him Congress would kill the plans the moment they arrived.

It was shortly before the holiday recess in December 1971 that the Subcommittee on Space Science and Applications at last got around to deciding the matter, assembling to vote on the proposed NASA budget for 1972. For weeks before that, Victoria Melikan had been on Capitol Hill lobbying the congressmen hard for the JPL budget in general and the TOPS project in particular. She'd practiced her arguments about how much there was to learn from the five planets; memorized the figures showing how much more expensive it would be to fly five different spacecraft to five different worlds instead of making one majestic sweep past the whole group; remembered especially to remind the congressmen that the outer planets' alignment wouldn't occur again until the year 2153 and it just wasn't scientifically sensible—it wasn't even scientifically *moral*—not to seize such an opportunity when it presented itself. That last argument was her trump card, and in the weeks she was lobbying the legislators, she always made it a point to pull it out, enjoying how the congressmen's faces subtly changed as a debate that until now had been all fiscal and political suddenly turned vaguely existential. If any of Victoria's pitches was going to change congressional minds, that would be the likeliest one to do it. Shortly before the vote on the project, Minnesota representative Joe Karth, the chairman of the subcommittee, pulled her aside for a private moment to let her know if it had.

"This TOPS thing . . . ," he began hesitantly.

"Mm-hmm," Victoria said, smiling as agreeably as she could.

"We've given it a lot of thought."

Victoria was silent.

"And there's no way, just no way we can justify spending that kind of money."

"You know it's five planets—" Victoria began.

"—for the price of one mission," Karth finished. "I know."

"You know it would cost much more to visit them all separately."

"It costs enough as it is."

"You know this is an opportunity humanity isn't going to get for another hundred seventy-six years."

"Then humanity will have to worry about it, then."

Karth looked at Victoria, saw her struggling for something else to say, and decided to make it easier for her. "It's not going to happen," he said simply. "It's just not." He walked away without another word.

The formal vote on the TOPS plan came a few minutes later, and Victoria, as was her custom, sat in on the subcommittee session until the decision was officially rendered. Then she returned to NASA's Washington headquarters, phoned Pickering, and told him what had happened. As always, she was impressed by the seemingly unperturbed way he took bad news; as always, he remembered to thank her for her call; and, as always, she suspected that no sooner had he hung up the phone than he was busy formulating other plans.

At his desk in Pasadena, Pickering was formulating indeed.

•   •   •

The problem with trying to undertake a mission like the grand tour is that no matter how hard engineers work to design it and congressmen work to derail it, the planets themselves don't really care. Even as the Subcommittee on Space Science and Applications was taking its vote and Victoria was making her call to Pickering, the great worlds of the outer solar system were already beginning to ease their vast bulks into position, preparing for an alignment that was now just a few years away. Humanity could either find a way to be there for the event or not, but the planets—heedless of the doings of a tiny species on a tiny world deep within the inner solar system—were going to line up regardless. What Pickering and his designers had to do now was figure out if there was some other way to fly a mission to those worlds without relying on the budget-busting ship they had just designed. The answer, they suspected, might be in a far simpler spacecraft they had first built long ago.

For all of the different ships JPL engineers had invented in the thirteen years the lab had been in business, most were descended from a single progenitor machine: the *Ranger* moonship. With its sturdy armature, its simple solar power system, and its easily adjustable architecture on which any number of cameras and sensors could be hung, *Ranger* was the least expensive and most flexible spacecraft the lab had ever developed. The *Mariner* ships that flew to Mars and Venus were descended directly from the *Ranger*s, though modified for their longer missions and different targets. The still-to-be-built *Viking* ships—two-part, piggyback spacecraft that would both orbit and land on Mars sometime in the mid-1970s—were, in turn, descended from the *Mariner*s.

In designing the TOPS ship, JPL had done away with the *Ranger* bloodline altogether, seeking to develop an entirely new species of spacecraft from the inside out. Now, with that new species extinguished before it could even be birthed, Pickering instructed the engineers to return once again to the ship that had succeeded for them so many times before and see if they couldn't reconfigure it for deep-space flight.

That reconfiguration turned out to be relatively easy. Beginning with a basic *Ranger-Mariner* skeleton, the engineers stripped away its useless solar panels and replaced them with their newly designed nuclear power system. In order to prevent the more delicate ship from being damaged by the waste heat the system gave off, they mounted the generator on an eight-foot boom that reached away from the main body of the spacecraft like a bony arm. On another limb on the opposite side they mounted a cluster of off-the-shelf instruments, including a camera, an ultraviolet spectrometer, a plasma detector, and a photopolarimeter. In the center, they mounted the same prodigious communications dish they had designed for the

TOPS spacecraft, but in the heart of the ship they did away with the powerful, five-brained STAR computer and replaced it with a simpler, double-processor machine.

The spacecraft that resulted from all this workbench cobbling may have been less reliable than the magnificent TOPS, and it was certainly less lovely—nothing that the engineers would have designed if their hands hadn't been forced by Congress. But when the engineers stepped back and looked at their ship, they had to admit that lamentable as the machine may have appeared, it certainly looked as if it could fly. Just where it would fly was another matter entirely.

It was clear from the resounding repudiation TOPS received from Congress that even with a more cost-effective spacecraft, the government was not about to put itself on the hook for a five-planet mission that would consume manpower and resources for more than a dozen years. The answer for the flight planners was to start eliminating planets. Pluto was the first to go. A flash-frozen ball barely two-thirds the size of Earth's own moon, Pluto had been a questionable target from the start. With the thirteen years it would take to complete the trip to the distant world, it seemed like an out-of-reach target, too. Neptune was a far bigger world and a far richer one in terms of chemistry and overall scientific promise. But the problem was, Neptune also lay at a breathtaking distance from Earth. Indeed, so far away was the eighth planet that for twenty years out of every 250 or so, it actually became the ninth planet, as the 3.7-billion-mile egg-shaped orbit Pluto inscribes around the sun caused it to cross inside Neptune's 2.8-billion-mile circular one. That twenty-year period was approaching at about the same time the outer planets mission would have to be launched, meaning that a flight to Neptune would take only about a year less than a flight to Pluto—clearly too many years for the new, less sturdy ship the JPL team had designed.

That left Jupiter, Saturn, and Uranus. Traveling at 35,000 or so miles per hour, the new spacecraft would need less than two years to reach Jupiter, which lies about 400 million miles from Earth, and little more than three years to reach Saturn, which lies about 800 million miles away. Uranus, however, is more than twice as far as

Saturn, orbiting nearly 1.7 billion miles out. Trudging such a great way through the solar system would take at least a good eight years—or almost seven years longer than any other spacecraft had ever had to fly to reach its target planet before.

The only prudent answer then was to pass up Uranus as well and pare the grand tour down to a simple, two-planet hop to Jupiter and Saturn. The JPL planners still hoped to send two ships to fly the mission, launching them on essentially the same flight path to ensure that at least one of them would get where it was going if the other should fail. Nonetheless, even with a two-ship fleet to build and launch, JPL estimated it could bring the entire project in at a relatively pennywise $250 million, or just a third of the price tag of its TOPS predecessor—clearly a more appealing prospect to budget-conscious legislators. In mid-1972, NASA submitted this scaled-down plan to Congress and this time Congress quickly approved it. In July the new project—which Congress began referring to by the clumsy moniker *Mariner* Jupiter-Saturn 1977, or MJS77—officially got under way on JPL's Pasadena campus.

At first, work on the MJS77 program began like work on any other JPL program. The engineers were given the spacecraft specs, the flight planners were given the planetary targets, and both went about the job of preparing to fly the mission NASA told them to fly. But at least a few of the designers and planners were contemplating something else. Sending the new spacecraft to Jupiter and Saturn was a fine shakedown cruise. But suppose the ships were still operating after that? Suppose they finished their work at Jupiter, finished their work at Saturn, and then found themselves fit and well, poised on the gravitational precipice between Saturn and the true cosmic deep? To be sure, that was a big supposition, what with the makeshift design of the spacecraft that would be flying the mission. But if by some chance the machines *were* still functioning after their 800-million-mile journey, did it make sense—any sense—simply to shut them down? It was one thing to refuse to spend the money to build a spacecraft capable of making the long-haul journey to Uranus, Neptune, and Pluto. It was another thing entirely to build a cheaper spacecraft that just happened to be able to fly farther than the spec sheets said, and then not push it as far as it could go.

Congress, clearly, had already said no to such a plan; but in the space game, there were noes and there were noes. And three years after launch, when the new ships were deep in space and had completed their prime mission, it was always possible that just enough cajoling and just enough wheedling could persuade Congress to change the no to a yes. The key for now was to shut up about any mission but the formally approved MJS77, stay within the $250 million budget, and be exquisitely careful about how the new spacecraft were designed. Nothing should be built into either ship that had an absolute life span of only three or so years. The two machines that reached Saturn around 1980 had to be as fit as possible for another eight or nine years in space so that if the green light was given for the planets beyond, they could simply point their prows and go.

Certainly, few of the engineers at JPL would speak openly about so seditious an idea. But in other, subtler ways the word got out. Bud Schurmeier, the engineer who had assumed the reins of the *Ranger* program when Jim Burke was sacked, had been named manager of MJS77, too. John Casani, another *Ranger* veteran who had been working on other projects at JPL, was named his lieutenant. Coming to work for Schurmeier meant Casani would have to change offices, and changing offices meant he would have to change his telephone extension. The day he moved in, he called the JPL operator and made a special request.

"I'm new to the MJS77 office," he said. "What's the chance of getting the number 6578 assigned to me?"

The operator flipped through her phone log. "It's pretty good," she said at length. "No one else is using it."

"Let me have it then," Casani instructed.

"It has to be 6578?" the operator asked as she made the notation.

"Precisely 6578."

Later that day, with his phone operating, Casani dropped by Schurmeier's office and handed him a slip of paper.

"New phone number," he said to his boss. "Like it?"

"It's a number," Schurmeier said with a shrug.

"Not just a number. Look at the paper and then look at your phone," Casani instructed. Schurmeier complied. "6578 is MJSU," Casani said with a broad smile. "We're going to Uranus, Bud."

Schurmeier dropped his head in his hands. "Jupiter and Saturn, John. I want you thinking only about Jupiter and Saturn."

"I will," Casani said cheerily. "For now."

•   •   •

Enthusiastic as the JPL engineers were about the idea of sending their spacecraft to as many planets as possible, they did not much care what the ships actually did when they got there. If the planetary scientists wanted to take pictures the engineers would build them a camera. If they wanted to study chemistry or magnetism they'd build them a spectrometer or a magnetometer. The engineers were in the business of designing hardware and assembling the ships; after that, it was up to the scientists to put them to use. The scientists themselves had very definite feelings about the matter and very specific plans for the spacecraft. And while they were intent on visiting and studying the outer planets, what they wanted to study especially closely were those planets' moons.

It had been a long time since scientists began to appreciate the curious appeal of the solar system's moons—more than 350 years, in fact. Unlike many of the pivotal moments in the near-timeless field of astronomy, the study of the moons could be traced to an extremely specific date: Thursday, January 7, 1610, at about one o'clock in the morning.

It was cold and clear in Padua, Italy, that night, one of many punishingly cold nights the city had experienced that winter. Cold was not so good if your work was stargazing, but clear was very good indeed. When the night broke cloudless and sparkling, Galileo Galilei, a tenured professor at the University of Padua, who had made it his business to do a lot of stargazing lately, carried his little two-foot-long telescope outside for as many hours of studying the sky as the frigid temperatures would allow.

Galileo's telescope was a fine one—lightweight, portable, and yet capable of magnifying objects a remarkable thirty times. To be su. it was not the first telescope in the known world; the first telescope or at least the first Galileo had heard of—had been developed the year before by a Dutch optician, Hans Lippershey. Living in faraway Padua, Galileo had never had a chance to examine Lippershey's instrument himself, but after reading a detailed account of it in a scientific journal, he was pretty sure he could design a better one. Before the year was out, he had indeed built a modest scope with a nine-power magnification that rivaled Lippershey's own.

When Galileo completed the instrument, he knew that this was not an achievement he should keep to himself and quickly invited the members of the Venetian Senate to come see what he'd built. So dazzled were the lawmakers when Galileo led them up a Paduan tower and showed them the swollen, swimming images of the local countryside that appeared in his lens that they quickly doubled his salary and made his provisional tenure at the local university a permanent one. With the time and wealth such good fortune brings, Galileo was able to devote himself to building his improved, thirty-power instrument. As he grew more skilled at using the new scope, he'd begun putting it to work to study not ships at sea and remote hills, but the bodies of the distant cosmos themselves. The cloudless night of January 7 struck Galileo as just the right opportunity for those kinds of observations.

As this evening's work began, Galileo first turned his attention to the planet Jupiter. The fifth of the six known planets had always been a seductive target for stargazers, mostly because of its great size and brilliance. On other nights, Galileo had pointed the maw of his telescope toward the remote world, framed its fuzzy, vaguely striated face in his narrow field of vision, and followed the planet as it swam through the stationary stars that surrounded it, wandering across the firmament from one evening to the next, even as the stars themselves hung fixed in the sky.

Most of those stars were of little interest to Galileo—at least on a night when he had set his sights on a planet. This evening, however, as he brought Jupiter into focus in his eyepiece, three unusual stellar pinpoints caught his eye. Unlike the other stars that surrounded gi-

.dom spangle, these three seemed almost
..net—two lying just to the east of it and one
..ore curiously, they were not located just anywhere
..ate vicinity of the world, but in what appeared to be a
..ine, queued up almost precisely along the same latitude as
planet's equator. Though the soupy Earthly atmosphere and
Galileo's own tired eyes made it impossible to be certain, it also appeared that the three points were not all the same size, with the west-
ern- and easternmost ones appearing slightly larger than the one
between them.

The queer new stars intrigued Galileo, but only mildly. Even
amidst the most chaotic disorder there could sometimes be little
pockets of order, and if the ancient constellations proved anything, it
was that, with an infinite number of random stars decorating the sky,
at least a few would occasionally position themselves in not so random ways.

The next night, however, when Galileo once again turned his
telescope to the sky, he noticed a different state of stellar affairs. The
three little stars were still visible, still lined up along the same plane
as Jupiter's equator, but tonight, all of them had somehow migrated
to the western side of the planet. This kind of motion was beyond
the powers of any known star and that left Galileo feeling uneasy.
Generations of astronomers before him had painstakingly calculated
the speed at which all of the planets change their position in the sky
from a particular hour one night to the same hour the next night.
From what Galileo was seeing now, their calculations of Jupiter's
speed were apparently wrong, with the planet moving far faster than
once thought—so fast, in fact, that in a span of just twenty-four
hours, it had left all three stars in its western wake. By tomorrow
night, Galileo figured, the three little points should be even further
behind.

But the little points weren't. During his next observation Galileo
noticed that one of his mysterious stars had inexplicably disappeared
altogether, while the remaining two had returned to Jupiter's eastern
side. The night after that, one of the two stars had moved to the
west; the following night, a third star reappeared on the east. The

Io   Springing volcanic leaks all over its scalded surface, Io may be the most explosively active world in the solar system. Gravitational pressure from Jupiter and its other satellites causes rhythmic pulsing deep within the moon, which in turn leads to volcanoes. This picture was taken by the *Galileo* spacecraft at a distance of 302,000 miles.

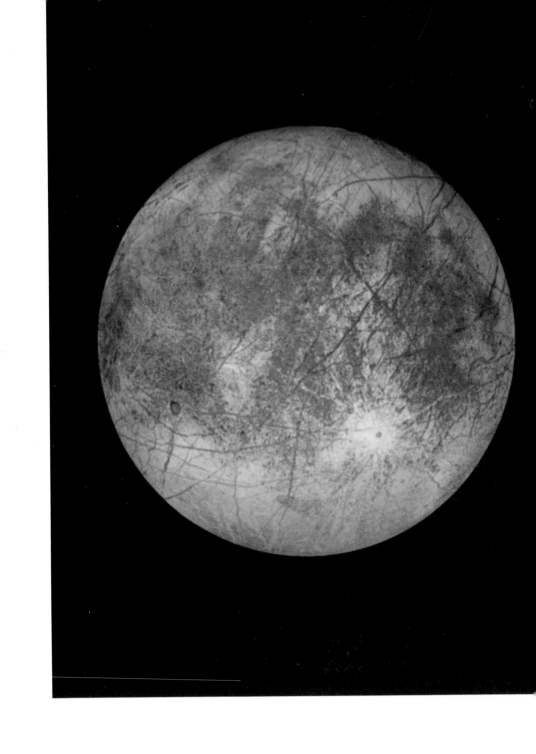

EUROPA   Europa's brittle, eggshell surface is made entirely of water ice. The absence of craters suggests that the ice regularly melts; the webwork of visible cracks suggests that something is surging not far below. Taken together, all of this points to a rich, briny, global ocean lying just beneath the surface—an ocean that could easily support Earth-like life. This picture was taken by *Galileo* from a distance of 417,900 miles. (The image at left is Europa's natural color; below, enhanced color highlights surface differences in the moon.)

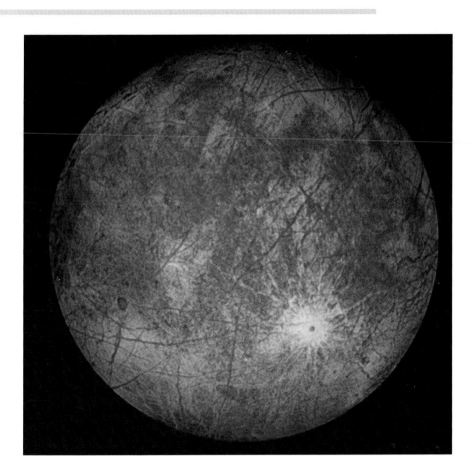

GANYMEDE   The biggest moon in the solar system, Ganymede has a diameter of 3,262 miles, making it larger even than some planets. A rice-pudding world composed of rocks and ice, Ganymede is covered by a crust made up of ancient, cratered terrain and smoother plains where the topography was refreshed by upwelling slush. This picture was taken by *Voyager 2* from a distance of 744,000 miles.

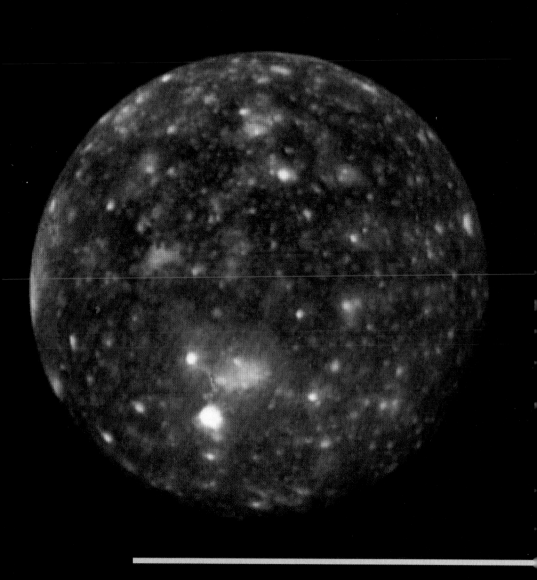

CALLISTO   The most distant—and the coldest—of Jupiter's four large moons, Callisto has remained virtually unchanged for the better part of 4.5 billion years. Its frozen face still carries the scars of nearly every meteorite that ever hit it, though over time, the contours of those craters have softened a bit as the surface ice has slowly slumped. This picture was taken by *Voyager 2* at a distance of 1.4 million miles.

TITAN   Saturn's largest moon, Titan is one of the most chemically rich places in the solar system. Drenched in ethane, methane, and other organics, it is thought to be a dead ringer for the prebiological Earth—if the prebiological Earth were somehow transported 800 million miles away and flash-frozen in deep space. *Voyager 1* flew by Titan at a distance of just 4,000 miles, but the moon's dense atmosphere prevented the spacecraft's cameras from seeing a thing.

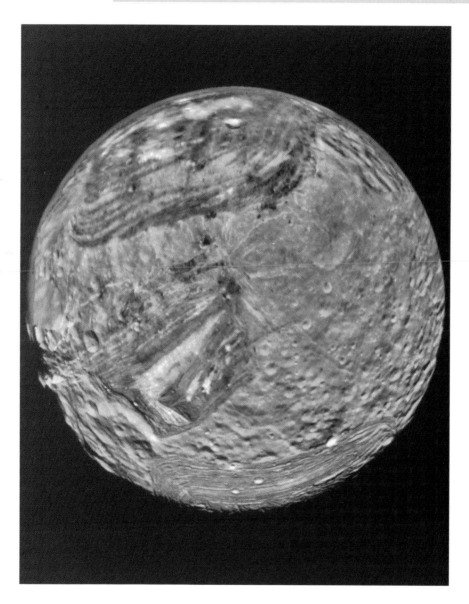

MIRANDA   Flying by at a distance of 19,000 miles, *Voyager 2* photographed the fractured face of Uranus's moon Miranda. Gravitational turbulence from surrounding bodies may have shattered Miranda to shrapnel. The
accumulated gravity of all those fragments, however, allowed the moon to reassemble itself into whatever shape best approximated a sphere.

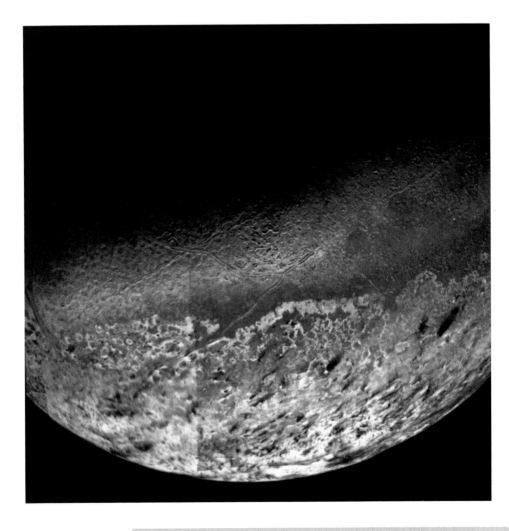

TRITON   At −391°F, Neptune's large moon Triton is the coldest place in the solar system. Its surface is made up of a permafrost of nitrogen, carbon dioxide, carbon monoxide, and methane—materials that would normally vaporize at the barest breath of solar warmth. Flying by just 15,500 miles above the moon, *Voyager 2* nonetheless noticed dark carbon geysers occasionally blasting through the ice, as the faint flickers of sunlight that reach Triton heat up subsurface materials and cause them to expand.

stars continued changing position and numbers until finally, on the ninth night of his observations, Galileo saw no fewer than four of the mysterious points, all in close proximity to Jupiter, all in a tidy rank along its equatorial waistline. Now, suddenly and simply, the astronomer understood what he was seeing.

These seeming stars weren't part of the stellar background canvas at all. They were wards of Jupiter, moons of Jupiter, as inextricably tied to the planet as if they had been *lashed* to Jupiter. The great, distant world did not hurry alone through the nighttime skies like Saturn, Mars, Venus, and speedy little Mercury. Rather, like Earth, it was forever accompanied on its journey; while Earth was circled by just one companion body, however, Jupiter was surrounded by four. When that evening began, Galileo—and, by scientific extension, humanity as a whole—knew of but one moon in the vast solar system. When that evening ended, the number had suddenly grown to five.

Coming up with names for the clutch of new moons was not a simple matter, not like naming the planets had been, anyway. It was the ancients—the Romans, the Greeks, the civilizations that came even earlier—who discovered the planets. In those less cynical, less clinical times, grand celestial bodies needed grand, heavenly names, so the planets were given the names of the gods.

But the first century was not the seventeenth, and in the era of science and finance, there were other, less exalted things to consider. It was not an accident that even before designing his telescope, Galileo had already distinguished himself as one of the great academicians of his land, having risen to his provisional and then permanent professorships. Such academic positions required influence, influence required powerful supporters, and Galileo knew how to attract both. Gifted in language, Galileo enjoyed writing, and did so with cunning versatility. When he was composing an essay for popular consumption, he generally wrote in the same slouchy vernacular he knew his readers used. When he was writing for other academics or potential benefactors, however, he would drop such studied plainspokenness in favor of the florid and floral Latin that was his natural tongue. If he had a controversial idea to introduce, an idea that was sure to inflame scholars or government officials, he would trick out his writing in even more decorative language than usual, pick a few of his likeliest critics and then—with gratitude and blandishments— dedicate the work to them. Their eminences generally appreciated the homage and often as not rewarded Galileo with their support and their applause.

It was no surprise, then, that when Galileo discovered his four new moons he should take the occasion to honor a group of people far more worthy of acclaim than he—a mere salaried professor— could ever hope to be. With a humbleness that surprised none of his peers, Galileo announced that the new worlds he had just found would henceforth be known as the Medicean Stars.

The Medicis, of course, for whom the moons were self-evidently named, were members of a powerful family of financiers and business leaders who had been ruling Florence and Tuscany since 1434. It was the Medicis who had long determined how local land was held, how wealth was distributed, how industries were run, and even how the occasional war was fought. And it was the Medicis, not incidentally, who determined who should fill the chairs and win the tenures at the country's great universities. In Padua, where Galileo lived, it was Cosimo de Medici, the grand duke of Tuscany, who handled such affairs, and when Galileo published his first book, *The Starry Messenger,* in 1610, making public the discovery of his Jovian moons, he did not shrink from explaining how he had chosen the names for the new bodies.

"... the maker of the stars himself has seemed by clear indications to direct that I assign these new planets Your Highness' famous name in preference to all others," he wrote. "For just as these stars, like children worthy of their sire, never leave the side of Jupiter by any appreciable distance, so as, indeed, who does not know clemency, kindness of hearth, gentleness of manner, splendor of royal blood, nobility in public affairs, and excellence of authority and rule have all fixed their abode and habitation in Your Highness." Cosimo de Medici himself saw little to quarrel with in this, and months after the new book appeared, decided that a scholar of Galileo's stature was wasting his considerable skills in an academic backwater like Padua.

From now on, Medici declared, the gifted astronomer would fill the chair of first professor of philosophy and mathematics at the University of Pisa.

Simply because a celestial name served the career of Galileo Galilei, however, did not mean it served the interests or satisfied the tastes of astronomers as a whole. Simon Marius, a German stargazer and righteous rival of the favor-seeking Galileo, was more dissatisfied than most. In 1606, Marius was studying at the University of Padua when he invented—or believed he invented or pretended he invented, depending on just who at the university was telling the story—a marvelous little calculating instrument called the proportional compass. No matter how Marius came by the instrument, he quickly went public with it. No sooner did he do so, however, than an outraged Galileo rose up to challenge him, insisting that he had dreamed up just such a device nine years earlier and that Marius had obviously filched the idea. Wielding far more influence at the university than Marius ever could, Galileo quickly had the insolent young student expelled on a charge of plagiarism and sent packing back to Germany.

Three years later, when Galileo began popularizing the use of the telescope, Marius, learning of the development, claimed that he himself first got a look at a simple telescope in Frankfurt in 1608 and that there was nothing new in Galileo's work. When Galileo announced that he had discovered his four Jovian moons in January 1610, Marius insisted that he had spotted them, too, and had done so in December 1609.

But the grim Marius was no match for the glad-handing Galileo, and it was Galileo who earned universal credit for finding the new satellites. Marius, however, still had an argument to make. Regardless of who discovered the bodies, was it right for the self-interested Galileo to determine what they would be called? The pandering evident in naming such magnificent worlds after a mere Tuscan industrialist demeaned the very moons themselves, Marius argued. He recommended something else entirely. Since the world around which the moons orbited was named in honor of the Greek's Zeus and the

Roman's Jupiter, its retinue of moons should justly be named after members of Jupiter's own court. The innermost satellite, he insisted, should be named Io, after one of Zeus's most beloved maidens; the second would be Europa, a Phoenician princess abducted by Zeus and later the mother of three of his children; the third would be Ganymede, a Trojan boy whom Zeus made cupbearer to the gods; the fourth and last would be Callisto, a nymph loved by Zeus.

Marius knew instinctively that astronomers would take to these names—certainly more readily than they had taken to Galileo's— and, as it turned out, he was right. No sooner had he floated his idea than scientists around Europe began writing, speaking, and teaching about the marvelous new worlds Io, Europa, Ganymede, and Callisto. With each such reference, the gulf between Galileo's lofty position in the scientific world and Marius's far more modest one closed a little bit. Galileo, after all, may have birthed the Jovian moons, but it was the lesser Marius who determined how they'd forever be known.

•   •   •

Over the years that followed Galileo's discovery, the business of spotting and naming moons picked up momentum. Before the seventeenth century was out, Holland's Christiaan Huygens found a vast, yellow-orange moon circling the already gaudily ringed planet Saturn. Huygens named his new moon Titan, after a family of giants in Greek mythology who once sought to rule the heavens but were overthrown by Zeus. Giovanni Cassini—a countryman of Galileo— followed with the discovery of four other Saturnian moons, which he named Iapetus, Rhea, Dione, and Tethys, after other mythical Greek figures. In the eighteenth century, the British astronomer William Herschel discovered the planet Uranus and quickly found Oberon and Titania—which he named after characters created by his beloved Shakespeare—circling it. He then found Enceladus and Mimas, two additional moons orbiting nearby Saturn. In the nineteenth century, other astronomers found the planet Neptune and its big moon Triton. After that came Hyperion, yet another moon for Saturn, and Ariel and Umbriel, two more for Uranus.

There was not much beyond the obvious that the astronomers could determine about any of these new satellites from the fantastic distance at which they were being observed. The worlds were obviously rather big, since how otherwise could they be seen? They were obviously round like Earth and the other planets, since what other shape could they be? More than that, the scientists couldn't say. So taken were they with the very fact of the satellites' discovery, however, that many of them didn't even *care* to say.

But even as the family of planets and moons was growing so steadily, something about the solar system puzzled astronomers— something that appeared to be missing from it entirely. In 1766, Johann Titius, a Prussian professor of astronomy, physics, and biology, had happened upon a troubling little equation. For no reason that he could determine, Titius was fooling about with numbers one day and found himself writing down a string of utterly meaningless figures— 0, 3, 6, 12, 24, 48, and 96—connected only by the tidy way the numbers doubled as they went along. Titius stared at the figures, then stared at them some more, and something faintly, faintly familiar seemed to tease at him—or something that *would* be familiar if he just did something to the numbers. Without knowing why, he realized that what he had to do was add to each one the numeral 4, the smallest whole number whose nonfractional square root is something other than the number itself. This changed Titius's pointless string of figures to a still pointless—and now far less tidy—4, 7, 10, 16, 28, 52, and 100. Finally, though again he did not know the reason, he divided those numbers by 10. The string of ugly fractionals he had left—.4, .7, 1, 1.6, 2.8, 5.2, and 10—had no meaning to him at all.

Or at first they didn't. But as Titius looked at the numbers, he all at once realized that they weren't just an arbitrary string of digits at all, but a ratio. And they weren't just any ratio; they were a planetary ratio. If the third number, the 1, were assigned to Earth, the first number would become Mercury, which Titius and other astronomers had calculated was .4 times as far from the sun as Earth is. The second number was Venus, which was .7 times Earth's distance. The fourth number was Mars, which was 1.6 times as far as Earth; later came Jupiter, at 5.2 times and Saturn at 10 times.

What struck Titius most wasn't just the sweet neatness of the astronomical coincidence; rather, what struck him was that the figures made gravitational sense. As the planets were coalescing out of the great cosmic windstorm that gave birth to the solar system, they couldn't pop up just anywhere. Form too close to a neighboring planet and the gravity of the two worlds would cause them to take hold of one another, drawing them together into a single megaworld. In order to survive as a discrete body, each fetal planet had to keep its distance from the next, with the precise distance being determined principally by the size of the planets on either side of it. In the case of the solar system as it was known so far, the balance wheels of gravity caused the planets to fall into orbital slots that precisely corresponded to Titius's seemingly random string of numbers.

But the equation had a flaw. In Titius's numeric sequence, there was one orphan number, his 2.8, that had no planet associated with it. If the formula was to be believed, somewhere in space, about 260 million miles from the sun—gravitationally centered between Mars and Jupiter—a world as real as any other by rights ought to be orbiting. But when centuries of astronomers looked in the direction of that vast stretch of cosmic real estate, empty space was all they ever saw. Things grew even more perplexing several years later when William Herschel turned his telescope toward the deep sky and found Uranus, just over nineteen times as far from the sun as Earth is—or exactly where the gravitational formula said it should be. Titius's orderly equation had now gotten more orderly still at its far end while remaining stubbornly corrupt in the middle.

That was how the equation stayed for the next thirty-five years, the mystery outliving even Titius himself. It was only on January 1, 1801—the first night of a new year in a new century—that the puzzle was solved. On that evening, Giuseppe Piazzi, a Theatine monk and a professor of mathematics at the Academy of Palermo, was conducting some telescopic observations of his own when he discovered a small, nearly insectile flicker somewhere between Mars and

Jupiter. Piazzi knew he couldn't be looking at a planet, since planets were great, hulking spherical things made out of gas or rock. What he was looking at could be no bigger than 125 or so miles across, a scrap of nothing that, given its small size, might not even be properly round.

If the body wasn't a planet, the only thing it could be was a comet. But when Piazzi tried to track its path, he found it didn't move like a comet at all. Comets typically careen through the solar system at a drunkenly inclined angle, cutting heedlessly across the flat orbital plane as they dive in toward the sun. Piazzi's putative comet moved in a much more orderly way than that, plodding in what appeared to be a tidy solar circle that was somewhat wider and slower than Mars's orbit and somewhat smaller and faster than Jupiter's.

Piazzi wasted little time going public with his mysterious new body, trusting that the rest of the astronomical brotherhood would help make sense of its meaning. The rest of the astronomical brotherhood quickly did just that. Though Piazzi's rock might be too small to be a planet, the other scientists concluded, the fact that it was orbiting the sun precisely where Titius's equation said a big body should be suggested that it might well be one of the bones of a planet. Long ago, it now seemed possible, a protoplanet might have begun forming in the Mars-Jupiter void, only to be shattered by a collision or some other cataclysm. Tumbling through so gravitationally complex a region, the rocky detritus that resulted might never have been able to reassemble itself into a world, and instead would simply orbit the sun as a permanent debris stream—a former planet, a mortal planet, reduced by violence to nothing but rubble.

That was the theory anyway, but before it could be confirmed, additional bits of the lost world had to be found. Over the next six years several of them were, as astronomers turned up more of these "asteroids"—or star-like bodies—in the Mars-Jupiter gap, none more than a few dozen miles across. The collected rubble still didn't amount to much, certainly not enough to form a full-sized planet, but if a few big bits of rock were turning up, hundreds, perhaps thousands, of smaller ones were almost certainly there, too.

Just how many asteroids there were and just how big a planet they once made up was impossible to say. Indeed, even if observers on Earth could somehow count every rock and pebble floating in the Mars-Jupiter rubble stream, they would still have a hard time making such a calculation, mostly because it eventually became clear that bodies orbiting in the asteroid belt didn't have to *stay* in the asteroid belt. Late one night in August 1877, a full human lifetime after Piazzi spotted the first asteroid, Asaph Hall, an American astronomer working at the United States Naval Observatory in Washington, D.C., was observing the planet Mars with the observatory's stout, twenty-six-inch-diameter telescope. Hall had picked a bad night for skygazing, with the air hanging thick and stewy in the former swampland that abutted the Potomac. Nonetheless, the sky was sufficiently free of clouds that when the steamy atmosphere occasionally held still, he was able to get an unusually sharp view of the planet, which was at that time an uncommonly close 53 million miles from Earth.

Bringing his instrument into focus, Hall began studying the familiar Martian disk with its ruddy color and its darker dapplings, when all at once he was brought up short by something not so familiar. Hovering directly by the world were two vanishingly small objects that appeared to be orbiting it. It was impossible to determine the size of the tiny satellites, but they couldn't be more than ten or twenty miles across, small enough not to be true spherical moons at all, but mere jagged boulders. Shards that small probably had not simply formed alongside their parent world the way the big, properly round moons had. Rather, they were almost certainly asteroids, refugee rocks from the nearby rubble belt that had somehow been knocked into free space by a collision of some kind and then been grabbed up and reeled in by the gravity of Mars. Once part of a true planet, the twin rocks had long ago been broken down to mere asteroids when the world they helped make up had fallen apart. Now, thanks to a few gravitational ricochets and a lucky catch by Mars, they had been upgraded again, orbiting a new planet and earning the status of fully certified moons. Hall named his tiny satellites Phobos and Deimos—or fear and panic—the mythical sons of Mars.

Hall's moons may have been modest things, but their meaning was enormous. If the silhouettes of Phobos and Deimos were so different from those of the other moons that circled the larger planets, their chemistry and history were likely to be different as well. And if huge moons and tiny moons could exist in the same solar system, who knew how many other species of satellites lay between them? Over the next century astronomers began to find out, as other big moons, other tiny moons, and litters of modest midsized moons tumbled through the telescopes. By the second half of the twentieth century, the solar system's satellite census had risen to a robust thirty-four, with one moon circling Earth, two circling Mars, thirteen orbiting Jupiter, eleven orbiting Saturn, five orbiting Uranus, and two orbiting distant Neptune. In the summer of 1977, a hundred years nearly to the day after the discovery of the Martian moons, a team of scientists in a scabby patch of desert in the southern half of California were preparing to launch a pair of spacecraft to visit most of those curious satellites and discover untold others that may lie among them.

4

Fire and Ice

Late Summer 1977

A lot had changed around the Jet Propulsion Laboratory in the five years since the MJS77 program had gotten under way. For one thing the MJS77 program didn't exist anymore. Oh, the spacecraft JPL hoped to launch to the deep solar system were still being built all right, and they were still being launched to Jupiter, Saturn, and who knew where else. But nobody had called the program by the name MJS77 in a long, long time.

Almost from the moment the *Mariner* Jupiter-Saturn 1977 mission was given the congressionally mandated moniker it was, most people around JPL figured it would have to change. First of all, the name was a mouthful. JPL mission planners had long made it a point to give their ships only clean, uncluttered, single-word names—*Explorer, Ranger, Mariner, Viking*—names as simple and stripped-to-the-chassis as the vehicles themselves. This one, however, was a polysyllabic mess.

For another thing, the name revealed too much. While most of JPL's projects had to be approved by administrators and legislators, it was generally the engineers themselves who called all the technical shots, choosing the places they wanted to fly and designing the spacecraft that would take them there. MJS77 had been a different story. From the outset, lawmakers in Washington had had their fingers in this particular mission's pie, using their budget-writing power to pick its destinations, define its objectives, and even dictate the kind of hardware that would make the trip. Give a group of legislators a chance to plan a project and they'll almost certainly want to name it. Give them a chance to name it and they'll almost certainly bungle the job, jamming as much information as they can into the title to prove that the money they're allocating will be well spent. A name like *Mariner* Jupiter-Saturn 1977 might have told you the model of the spacecraft, the places it was going, and the year it was scheduled to leave, but it also told you how the ship came to be— and in this case it clearly came to be in Washington, just another public works project brokered in the halls of Congress and approved in its committee rooms. That, the planetary explorers of JPL figured, was not the way taxpayers liked to think of their spaceships.

Shortly after the mission was approved, John Casani, the deputy project manager, held a competition among the JPL staff to see if it wasn't possible to choose a less unlovely handle for the new spacecraft. To his disappointment, however, the engineers came up empty. *Nomad* was a popular choice, but nobody especially cared for the vague whiff of aimlessness the name carried. *Pilgrim* was another favorite, but it too had a meandering feel about it. What Casani and his senior project managers really wanted to call the ship was *Voyager,* but they wondered whether they dared.

Several months before the grand tour was originally proposed, NASA and JPL had been considering another, equally ambitious mission to nearby Mars. Under that plan, the engineers would stack a pair of interplanetary spacecraft piggyback-style inside a Saturn-class booster and launch them into space together, targeting them to explore entirely different areas of the Martian surface simultaneously. The project, which had tentatively been dubbed *Voyager,* had an attractive simplicity about it, but ultimately it proved to be *too* simple. Launching two spacecraft atop just one rocket meant that if that rocket should blow during launch, both ships would be lost instead of just one. For a space agency that lived and died by the credo of hardware redundancy, that kind of twin-ship disaster was unacceptable. Eventually, nervous NASA officials scrapped the *Voyager* project, and when they did, they scrapped the name along with it.

Now, there was another mission getting ready to fly—another

mission in need of a swashbuckling name—and the orphaned *Voyager* seemed like a natural. The JPL bosses, however, who ordinarily were not a suspicious lot, were uncharacteristically skittish about giving their ships a borrowed name from a junked spacecraft that hadn't even been able to get itself off the drafting board and out to the launch pad, much less catapult itself into space. Every time the *Voyager* name came up for consideration, the members of the MJS77 team would fall into an awkward silence, knowing *Voyager* was a perfectly good choice, but knowing, too, that it was a mystically marked one. After a few awkward moments, somebody would come up with a reason to table the decision and the scientists would move on to other matters.

Eventually, however, with the project approved and the space agency press machine preparing for a big publicity push, the NASA bosses began insisting that the project managers pick *something* and be done with it. With nothing else remotely appealing—and none of the JPL staffers willing to admit why they were refusing to use the perfectly suitable *Voyager*—the engineers threw up their hands. There would be two ships built and launched on this outrageous mission to the outer planets, and for better or worse, they would henceforth be known as *Voyager 1* and *Voyager 2*. If fate was going to punish that choice, that was simply a risk the engineers would have to take.

Undergoing an even more important change than the name of the ships was the trajectory they would fly. Of all the planets and moons the twin spacecraft would be sent to visit, the one that intrigued the scientists the most was Titan, the largest satellite of the planet Saturn. At 3,193 miles in diameter, Titan is the second biggest moon in the solar system—after Jupiter's Ganymede—and is, in fact, bigger than two fully certified planets: Mercury and Pluto. What's more, Titan was thought to be an extremely chemically complex moon, dense with ethane, methane, and other hydrocarbons essential to life. A world this big with this kind of biological potential clearly warranted a good, close look. The problem was, actually getting to Titan was not going to be an easy matter.

In order to make the most thorough study of the moon possible, it wouldn't suffice for the *Voyager*s merely to fly by it on a horizontal trajectory that would take them past Titan's equator. Rather, it was necessary to make a steep, ascending approach, climbing vertically alongside the world in a trajectory that would allow a spacecraft to soar over both Titan's south and north hemispheres, while at the same time getting a glimpse of at least parts of its eastern and western halves as the moon rotated obligingly beneath it.

A *Voyager* attempting to make such a sharp south-to-north turn over a moon could not rely on its comparatively feeble engines alone. Instead, it would have to fly below the south pole of Titan and allow the gravity of the moon to grab it and flick it northward, helpfully tossing it along just the trajectory the JPL planners wanted it to follow. As powerful as Titan's gravitational throw would be, it would be impossible for the spacecraft to reverse course after its flyby of the moon was over and resume its horizontal trajectory past the planets. Instead, it would simply soar up and out of the solar system, sailing off forever into empty space. For a ship that was only going to visit Jupiter and Saturn, this would not be such a bad thing. For a ship whose builders were still harboring secret dreams of continuing on to Uranus and Neptune, however, this meant trouble.

For this reason, the JPL planners decided to send *Voyagers 1* and *2* on two different trajectories. *Voyager 1*, it was agreed, would be launched on a relatively speedy trajectory that would take it to Jupiter and Saturn, with special emphasis on the close Titan flyby, after which it would conclude its mission. *Voyager 2* would similarly be targeted for both Jupiter and Saturn, but would fly a slower route that would carry it past both worlds in the flat, allowing it to maintain a horizontal trajectory that would give it a far poorer look at Titan but at least a fighting chance at Uranus and Neptune. If *Voyager 1* for some reason failed in its Titan flyby, *Voyager 2* would be retargeted en route to make a second attempt at the giant moon. Such a course change would, of course, send *Voyager 2* out of the solar system, too, ruling out a visit to any worlds beyond Saturn. But so alluring was the biologic promise of Titan that the JPL scientists figured it was worth the risk.

Even as the *Voyager* mission's flight plan was changing, its trailblazing status was, too. Ever since the grand tour was proposed in 1966, JPL had been boasting that its ships would not only explore the outer solar system, but that they would be the *first* to do so, flying out to Jupiter and Saturn and perhaps Uranus and Neptune before either the Russians or any other NASA lab had even made it beyond Mars. But no sooner was the mission approved and the funding in place than two other spacecraft beat JPL to the punch.

In March 1972 and April 1973—years before the *Voyagers* were even scheduled to leave the ground—the aptly named *Pioneer 10* and *Pioneer 11* blasted off the pad in Cape Canaveral and into space toward their own planned rendezvous with Jupiter and Saturn. The ships were not designed and built by one of the grand NASA facilities in Houston or Florida or Pasadena, but rather, by the comparatively obscure Ames Research Center in Moffett Field, California. Ames, like JPL, was an old facility, having been in operation for years before joining the NASA family in 1958. Unlike JPL, however, Ames did not begin its life as an academic institution, but rather as a military one, serving as a research base and a training ground where new airplanes were developed and tested.

With its more modest scientific credentials, Ames was never considered one of NASA's frontline labs and was rarely given the glamorous mission-planning and spacecraft-building jobs JPL was. Instead, the Ames scientists were assigned the lower-profile job of conducting basic research and testing basic hardware, helping to support the more celebrated NASA facilities that were busy doing their more celebrated work. It thus came as something of a surprise in the early 1970s when Ames announced that it was planning to strike out on its own, building a pair of brand-new spacecraft and flying them to the same two planets that were the *Voyager* mission's prime targets.

The ambition—to say nothing of the presumptuousness—of the project took the space community by surprise. For all its derring-do, however, the *Pioneer* project was a relatively simple one, mostly because the spacecraft themselves were simple, too. The ships Ames was building were spindly little things, weighing barely 570 pounds apiece. Lightweight ships could carry only lightweight hardware, and the *Pioneer*s would not be carrying much at all—a camera, a few Geiger counter–like sensors, a radio, and a computer. A putt-putt spacecraft like this could be built on the cheap, and when the Ames researchers presented their plan to NASA and Congress, the administrators and lawmakers quickly approved it. If the *Pioneer*s failed, there'd be little in the way of resources lost; if by chance they succeeded, much would be gained scientifically. Meanwhile, JPL could continue working on its own, more ambitious project scheduled for later in the decade.

Though most of the engineers at JPL wouldn't admit it, they were only too happy to let Ames be the first to dip a toe into deep space. Traveling to the outer solar system meant traveling through the asteroid belt, a shooting gallery of flying rocks that could easily pulverize any ship that tried to pass through it. No spacecraft had yet attempted the tricky piloting that would be necessary to negotiate this region, and if the Ames planners were volunteering to run a little reconnaissance, the JPL planners would not argue.

Just as menacing as the environment between Mars and Jupiter was the environment in the immediate vicinity of Jupiter itself. Astronomers had always suspected that a world as big as Jupiter emitted a powerful—even lethal—radiation field, one that could poison a spacecraft's instruments before it could take so much as a single reading. No one knew exactly how toxic the radiation field was, but once again, if Ames was willing to be the canary going down into that particular coal mine, JPL would not stand in its way.

When *Pioneer*s 10 and 11, against all predictions, were successfully launched, and then, against all odds, actually arrived at Jupiter in 1973 and 1974, the news they sent back was both good and bad. The asteroid belt, it appeared, was a navigable place. When the ships reached that region of the solar system, they found that the population of rocks was even greater than expected, but the stretch of space in which they moved was so vast that an object as small as a spacecraft could pick its way through with little danger.

If the asteroid belt was less perilous than feared, however, the radiation field around Jupiter was far more so. Even when the two ships were nearly four million miles from the giant planet, their

Geiger counters began to twitch off the high end of the scale. The closer they got, the more punishment they absorbed, until the detectors themselves, as well as several other minor instruments aboard the spacecraft, became saturated and failed. At the ships' closest approach to Jupiter they were absorbing a radiation dose of about 450,000 rads, or about a thousand times the 450-rad level that is usually considered fatal for a human being.

After the ships' sizzling encounters with Jupiter, *Pioneer 10* was steered out of the solar system altogether, and *Pioneer 11*—or what was left of it—was sent on a lazy trajectory to Saturn that would allow it to reach the planet by sometime in late 1979. Ames, abiding by the compulsory esprit de corps NASA expected of its labs, shared all of its Jupiter findings with the scientists at JPL, who thanked their fellow engineers for their data, and then promptly put the information to use, adjusting their asteroid-belt flight path to follow *Pioneer*'s own, and reinforcing their spacecraft with radiation shielding that would prevent the ships from growing sick the way Ames's had.

The most significant change of all around JPL in the last few years, however, concerned not the spacecraft the lab built, but the people who built them. In 1976, William Pickering at last retired from the Pasadena lab he helped establish, deciding to go to work in the oil industry and, for the first time, earn the kind of private sector money his long-ago Caltech education entitled him to make. Stepping down at the same time was Bud Schurmeier. Chosen to head up the MJS77 flights almost the moment they were approved in 1972, Schurmeier had been working steadily on the project ever since. Half a decade was generally considered more than enough time for any one manager to oversee any one mission, and not long before the first scheduled launch of the first *Voyager,* Schurmeier was rotated out of the manager's spot, assigned to other JPL projects, and replaced by his assistant, John Casani. The cautious and conservative Schurmeier had been just the right man to steer the *Voyager* project from its planning stage through its development stage and up to the moment of launch. Casani, who from the very beginning had wanted to take the whip hand to the *Voyager*s and gallop them as deep into the solar system as their off-the-shelf parts could take them, would be just the right man to send them to the planets.

On August 20, 1977, at 8:56:01 A.M. Cape Canaveral time, the first *Voyager* interplanetary probe left its launch pad on the east coast of Florida and soared into space, carried atop a 159-foot Titan-Centaur booster. On September 5, 1977, at 10:29:45 A.M., the second Voyager followed. It took only minutes for the ships to reach Earth orbit and only an hour more for them to leave that safe, circular path and head out toward the stars. The journeys ahead of them would last years.

•   •   •

Scientists didn't know much about the moons of Jupiter in the months before the pair of *Voyager* spacecraft arrived at the planet. They knew their names, they knew their approximate size, and they knew—more or less—where to find them. For researchers hoping to take the census of the entire solar system, that wasn't much, but such fragmentary data at least gave them a place to begin their studies.

Io, Europa, Ganymede, and Callisto—the four moons Galileo had discovered 367 years before the *Voyager* launches—were by far the largest of the Jovian satellites. Measuring from 1,945 to 3,262 miles in diameter, they stay relatively close to the planet, flying in circular orbits that range from a low of 261,000 miles for Io to a high of 1.2 million miles for Callisto.

Far closer to Jupiter is tiny Amalthea. An irregularly shaped moon discovered in 1892, Amalthea measures only 117 miles across and flies at a cloudtop-grazing altitude of just 112,000 miles. Smaller and much more distant is an additional cluster of four tiny moons—Leda, Himalia, Lysithea, and Elara—discovered by different astronomers at different times throughout the twentieth century. Orbiting in reasonably close proximity to one another about 7 million miles above the surface, the moons measure from just 10 miles in width to a somewhat more substantial 115 miles. Farther still—some 15 million miles above the Jovian surface—is another four-moon cluster, Ananke, Carme, Pasiphae, and Sinope, also discovered one at a time throughout the twentieth century and also orbiting relatively close to one another. These four moons are even tinier than the lower cluster—measuring between 19 and 31 miles across—and circle the planet in a highly eccentric orbit, inclined about 105 degrees from Jupiter's equator. In addition to their sharp angle, the orbits of Ananke, Carme, Pasiphae, and Sinope are backward, following a clockwise path around the planet, unlike virtually all of the other bodies in the solar system, which orbit counterclockwise. How many other undiscovered moons there might be swarming among Jupiter's known litter of thirteen was impossible to say.

For *Voyager* scientists hoping to explore the Jovian system, the answers to these kinds of questions would come from the outside in. Approaching the planet and its moons would be like flying into a great metropolis. The spacecraft would pass the planetary provinces first, the deep-space wilderness where little was evident of Jupiter but the lines of magnetism and the waves of charged particles it gave off—the smokestack exhaust of a world still too far away to see clearly. Next would come the fiefs and duchies of the smaller outlying moons, then the hulking Galilean moons, and last the great body of the planet itself.

During the first week of 1979, close to a year and a half after the *Voyager*s had been launched, *Voyager 1* drew within 40 million miles of Jupiter, a distance at which the planet at last swam milkily into view. On January 4 the signal went up to the ship ordering it to open its long-shuttered eyes, switch on a pair of high-resolution video lenses—one built for close-ups, one built for panoramic shots—and beam back to Earth an image of what it was seeing. At first, what it saw was not much at all—a vaguely orange saccharin tablet of a world in which cloud patterns were visible, but just faintly; big moons were visible, but fainter still; and small moons were not discernible at all.

At just 40 million miles from the planet, however, *Voyager 1* needed to travel only another 20 million miles—a tiny fraction of the nearly 400 million miles it had covered so far—to halve its distance and double its resolution; at 20 million miles it had to travel just 10 million to halve it again; at 10 million miles another 5 million would do it. In steadily dwindling increments, therefore, the ship's eyesight grew better and better, and the improvements came faster and faster.

Even before the quality of the pictures the spacecraft could return had sharpened appreciably, the number it could return was already impressive. With its high-performance cameras gathering even the faintest shimmer of Jovian light and its cannon of an antenna blasting data back home, the ship was capable of taking and transmitting one complete picture every forty-eight seconds. This meant a lot of pictures indeed in the four months *Voyager 1* would spend in camera range of the Jupiter system—two months shooting straight on as it approached the planet and two months looking back over its shoulder as it passed.

As the pictures streamed in during those initial two months, the atmosphere changed palpably at JPL. The conference rooms and other common areas across the sprawling lab had long ago been fitted with closed-circuit televisions hung from the ceilings in heavy metal brackets. Most of the year these monitors carried nothing of much interest—time and weather readouts usually; perhaps a status report of an upcoming launch. Now, however, the screens fairly shimmered. From early in the morning to deep in the night, ever bigger, ever crisper pictures of Jupiter and its marble bag of multi-colored moons flashed and flashed and flashed again. In hallways and reception areas, knots of scientists and visitors would gather to watch as the newest images resolved themselves on the monitors. In the employee cafeteria, lines slowed and food service stopped as screens that only moments before had been sizzling with static suddenly began to flicker to life with the latest image of the giant planet 400 million miles away.

Though the *Voyagers* were designed to send home a wealth of other scientific data from the planets they visited, it was, not surprisingly, these cosmic portraits that caused the greatest sensation around JPL. And it was, also unsurprisingly, the technicians from the imaging team—the people who received the incoming signals from the ships and helped transform them from a mere electromagnetic data stream to crisp and comprehensible pictures—who earned the most applause from their Pasadena colleagues. Just as interested in the pictures that were coming back from the *Voyagers*, however, was another group of JPL investigators: the obscure and uncelebrated members of the *Voyager* navigation team.

There were few jobs in the entire JPL organization that were seen as less glamorous than the ones held by the scientists on the navigation team, who spent their days studying sky maps, calculating trajectory arcs, and scrutinizing hundreds, often thousands, of numbingly similar pictures the spacecraft took of star fields, in an attempt to confirm that the ships were pointed the right way. The environment in which this grindingly tedious work was done was spartan at best. With the exception of a few section chiefs and overseers, nearly all of the *Voyager* navigators were assigned to one end of the second floor of JPL's Building 264, a warren of open offices and cubicles tossed together in an overloud, overlit, utterly public jumble.

When the *Voyager* images were returning to Earth, the buzz in this noisy corner of the lab often grew overwhelming. Daily—sometimes it seemed hourly—the grand investigators from the grand imaging team would arrive with another stack of pictures for another busy navigator, wondering if it might be possible to chart this or that course or this or that coordinate from the sleet of static that flecked most of the pictures.

One day during the Jupiter phase of the *Voyager* missions, Steve Synnott, a member of the navigation team, was laboring over some images when he was approached with a slightly different request by Dave Jewitt and Ed Danielson, two Caltech researchers who were serving as consultants on the *Voyager* project. Earlier in the day Jewitt and Danielson had been studying a badly overexposed *Voyager* image and had noticed something strange. When a picture was overexposed, it meant that the camera shutter had been left open too long, causing things to run and blur. In the background of this particular picture, things had blurred indeed, so much so that the pinpoint-like stars the camera caught in its frame had all turned into tracer-like streaks. This in itself was not unexpected, but there was a detail in the picture that the imaging scientists didn't anticipate: One of the streaks, it appeared, was moving at an angle that was completely different from the rest of the star field. Since all stars remain fixed relative to a moving spacecraft, they should distort in the same direction. The fact that one was diverging from the others meant it couldn't be a star at all.

What Jewitt and Danielson wanted to know was what it was. The least exciting possibility was that the streak was nothing truly meaningful—a stray asteroid moving through space, perhaps, or an imaging glitch in the JPL computers. In the alternative, however, it might be something decidedly more important: a tiny, new, 10-mile-wide moon to add to Jupiter's already considerable brood.

If the object was indeed a moon, it ought to be in a calculable orbit, but there wasn't much in the image that would allow even a seasoned astronomer to determine what that orbit could be. The entire streak covered only about 40 pixels on the 800-pixel screen, and the entire screen represented only about 5 percent of the theorized moon's theorized orbit. Would it be possible, Jewitt and Danielson wondered, for Synnott to extrapolate the angle, altitude, and shape of the satellite's path around the planet from just that scrap of data and determine if it was in fact a satellite?

Synnott acknowledged that yes, such calculations might indeed be possible and yes he'd be willing to try. The job, he knew, would be a thankless one, since even if he did confirm Jewitt and Danielson's moon, the Caltech researchers would be the ones credited with its discovery. Nonetheless, he took on the assignment, deciding to tackle it only on weekends, when it wouldn't interfere with his other, more pressing work.

One Sunday evening not long after he first received the picture, Synnott was sitting at his desk in the uncharacteristically quiet navigation department, bent over a stack of Jupiter pictures, looking for more evidence of Jewitt and Danielson's maybe-moon. Giving one particularly vivid image a quick scan and preparing to slide it off his still-to-be-studied pile, he noticed something that brought him up short. There, at the edge of the image, small enough to be overlooked if you weren't trained in the business of *not* overlooking such things, was a pair of small black specks, ones that were nowhere near the likely location of the fleck he was looking for. The specks were situated directly over the central girth of Jupiter, directly over its equator, in fact, in a spot where the swirly orange-red mottling of the planet's atmosphere gave way to a cleaner all-orange band, making it far easier to pick out an object that might be flying overhead. Synnott snatched up another image and the dots were there again, still in lockstep with one another, but now in a different position relative to the cloud formations. A third picture revealed the same thing, but this time the objects had moved further still.

Synnott gulped, gaped, and had a sudden impulse to stand up—which he did for no reason at all. He then sat down again and smiled broadly. What he was seeing, he knew, was not an image of two bodies at all, but one: a Jovian moon sailing over the planet and casting its shadow on the cloudtops beneath it. That moon and that shadow were moving fast, he could tell, tearing along in an orbit that was far brisker than the leisurely flow of the ever-changing atmosphere below. What's more, the body and its shadow were relatively big, anywhere from 60 to 120 miles across, far bigger than the 10-or-so-mile body Jewitt and Danielson believed they had found, and located in an entirely different patch of Jovian sky. If this was indeed a moon, it might well be a brand-new one, and if it was a new moon, it belonged to Synnott alone.

Before Synnott could claim the discovery, he had to make sure that the satellite he had just spotted was not one of Jupiter's thirteen existing ones. If it was, it would most likely be the little, low-flying Amalthea. This innermost of Jupiter's moons, discovered nearly ninety years earlier by astronomer Edward Barnard, was more or less the same size as Synnott's theorized moon and orbited, as nearly as Synnott could tell, at more or less the same altitude.

For the next two weeks, Synnott continued to find his moon in picture after picture returned by the *Voyagers*, calculating its size and orbit more and more precisely. As he did, he became more and more elated. This was not the 117-mile-wide Amalthea, sailing along in its 112,000-mile-high orbit; and it certainly wasn't the little pebble of a moon Jewitt and Danielson thought they had seen. This was a body sixty miles wide, sailing through an orbit roughly 138,000 miles up. This was going to be Synnott's moon after all.

Within the week Synnott had amassed all of the pictures and all of the data he needed to make his claim, and had set them aside, waiting for the next time Ed Stone, the chief project scientist for the *Voyager* program, made one of his frequent walking tours of the *Voyager* bullpens. When he did, Synnott flagged him over to his desk and handed him a typewritten draft of a one-paragraph letter he was about to send to the International Astronomical Union in Cambridge, Massachusetts. The IAU, as anyone in the astronomical community knew, was the group you contacted if you had a new heavenly body to announce, and a one-paragraph letter was how you announced it. While astronomers everywhere were aware of what such a letter should look like, few of them ever got the chance to write one themselves. Stone read Synnott's single paragraph wordlessly and then looked up at him with just a ghost of a smile.

"Do you know its orbital period?" Stone asked.

"About eighteen hours," Synnott answered, handing Stone a sheet of figures.

"Its size?"

"About sixty miles," he said, handing over a small stack of especially sharp photos.

"Altitude?"

"138,000."

Stone glanced at the letter again and gave the pictures and the page of calculations a practiced scan. After a long moment he looked up and allowed the ghost of a smile to bloom into a real one.

"Well," he said with a good-natured shrug, "it looks like you've found yourself a moon."

Later that day Synnott sent off his letter, and a short while later the IAU responded, congratulating him on his discovery and sending along a list of possible names for his moon. Synnott studied the list and, at length, chose the name Thebe, after an ancient mythical nymph. Before long he found that he had cause to consult the same list again. Wading through more *Voyager* images, he found another, more modest, 25-mile moon orbiting 79,360 miles below his newly christened Thebe. This newer, smaller satellite he chose to name Metis, after the first wife of Jupiter. Shortly after that, he at last confirmed the existence of Jewitt and Danielson's moon, a 12-mile pebble of a thing orbiting just 79,967 miles up. The two Caltech scientists named their single satellite Adrastea, after the daughter of Jupiter.

The moon total of the largest known planet in the cosmos had now risen to sixteen, and not long after, the two *Voyager* spacecraft that had discovered the new worlds glided out of the Jovian system forever. Before the ships were gone, however, Linda Morabito, one of Synnott's colleagues in the navigation department, made one final and explosive discovery. Sitting down with a stack of overexposed images like the ones Synnott had been studying, she noticed a huge and curious bulge rising over the horizon of the big moon Io. The bulge could not be a mountain, she knew—it was just too big for that. Nor could it be another moon hiding behind Io; a body that large would long since have been spotted from Earth. The bulge, she knew, could only be a volcano, a mammoth surface eruption sending a vast cloud of exhaust high into space. Other pictures revealed other bulges indicating other eruptions elsewhere on the surface—as many as nine at any one time. Io, a world that had long been thought to be inert as a stone, was, in fact, fairly exploding with volcanoes, the only body in the solar system other than Earth known to be geothermally active. Morabito might not have spotted a moon of her own, but as scientific discoveries went, hers was even more significant than anything even Steve Synnott had achieved. Learning that the Jovian system was a crowded one was one thing; learning that it was a geologically living one was something else entirely. Closer study of the *Voyager* images would reveal just how alive the Jupiter family as a whole might be.

·   ·   ·

When the *Voyager* scientists at last had the time to analyze their portfolio of Jupiter pictures, one of the first things that caught their attention was neither the planet itself nor any of its moons, but another, related formation: a fine set of rings. The seventeenth-century astronomers who first turned their telescopes toward Jupiter's sister world, Saturn, were struck as much by the size and color of the planet itself as by its vast ring system. The bands were broad, bright, ostentatious things, more than twice as wide as the diameter of Saturn proper. Indeed, so vast and opaque were the rings that they actually appeared to obliterate the very stars that twinkled behind them, clearing a blacked-out zone ahead of and behind Saturn as the hoop-skirted world moved from one place in the sky to another. Had Saturn's rings been more tenuous things—straggly formations that were closer to mere strands than bands—they might have been utterly invisible from Earth; when they passed in front of the stars, however, they still would have blotted at least some of them out. When twentieth-century astronomers turned their attention toward Uranus and Neptune, they noticed precisely this kind of star-snuffing—which they called occultation—and concluded that the supposedly naked worlds were probably decorated by extremely faint rings of their own. No matter how powerful the telescopes, however, nobody had found a trace of even the faintest bands around Jupiter.

Voyagers 1 and *2* changed all that. As the spacecraft closed in on the planet, the relatively bright sunlight pouring in from over the ships' shoulders illuminated the world and its moons but bleached out anything smaller and fainter. When the spacecraft moved past Jupiter, however—placing themselves in a position that allowed the planet to blot out the sun—and then looked back at where they had just been, the world looked entirely different. Surrounding the now-shadowed sphere was an eerie, backlit nimbus that shrouded the planet like a luminous cloud. And within that cloud was a distinct set of nested rings.

Unlike Saturn's rings, a dense river of rubble and ice, Jupiter's rings were hopelessly tenuous, little thicker than the airborne dust that appears in the shaft of light pouring through an uncurtained window. When that dust forms itself into a circle and begins to orbit a world with a 279,000-mile waistline, however, even so faint a formation can look dramatic indeed. The *Voyagers* discovered at least three—and perhaps four—distinct rings around Jupiter, the biggest orbiting more than 35,000 miles above the planet, the smallest practically grazing its cloudtops.

Astronomers could not be certain what was responsible for Jupiter's bands—and all other planetary rings for that matter—but for more than a century they had had a pretty good idea. In 1850,

French astronomer Edouard Roche marshaled what knowledge he had of gravity, orbital dynamics, and lunar distances, and devised a formula for moon formation that became known, unsurprisingly, as the Roche limit. Assuming a planet and a moon have the exact same densities, Roche calculated that the moon could never safely move any closer to the planet than a distance equal to about 2.4 times the planet's radius. Slip too far inside the limit and the planet would tear the moon apart and drag the resulting debris down to the surface. Slip less deeply inside the limit—just barely cross the gravitational line—and the planet would still pulverize the moon, but the rubble that remained would be able to stay aloft, dispersing around the planet and transforming itself into a ring. As the relative densities of the planet and moon changed, the Roche limit would move in or out accordingly, but the satellite-shattering principle behind it would remain the same. In the case of Jupiter, the tenuous, planet-hugging nature of the newly discovered rings suggested that they had been formed by a small, low-flying moon, one that was no different from all of the other small Jovian satellites except that it had allowed itself to be grabbed too tightly by Jupiter's gravity fist and had paid a dreadful price.

"The stronger a moon is, the closer it has to be for the Roche limit to break it apart," said Rich Terrile, a planetary scientist at JPL. "Apply enough gravitational pressure and you can overcome the strength of any material. A moon inside the Roche limit experiences just that kind of pressure, and once it breaks apart, it all at once finds itself living in a world in which it can't put itself back together. Planetary rings are sometimes thought of as nothing more than the grave markers of moons."

Simply because one of Jupiter's small satellites was murdered by its parent planet, however, did not mean the others couldn't somehow survive, and plenty of them did. Of all of the undersized satellites, it was the innermost group of four—Metis, Adrastea, Amalthea, and Thebe—that intrigued scientists the most, particularly when they tried to figure out how the tiny satellites came to be there at all. The best guess for most of the astronomers was that they came from the asteroid belt between Mars and Jupiter. The more time scientists spent studying the giant debris field, the more they came to appreciate just how mobile the rocks that populate it could be. In addition to Phobos and Deimos, the two small Martian moons that appeared to have migrated out of the asteroid belt, there were at least two other clusters of belt rubble that seemed to have drifted away from the main stream. Those asteroids—which astronomers had dubbed the Trojan asteroids—had not gone into orbit around a planet the way Phobos and Deimos had, but into orbit along *with* it, taking up positions precisely 60 degrees in front of Jupiter and 60 degrees behind it and constantly maintaining those distances as the giant world inscribes its own orbit around the sun.

If both the Trojan asteroids and the Martian moons had come from the Mars-Jupiter debris belt, there was no reason that Metis, Adrastea, Amalthea, and Thebe—with their asteroid-like size and shape—wouldn't have originated there as well. If that was indeed their source, they could be carrying with them some revealing cosmic chemistry.

Meteorites falling to Earth are themselves often assumed to have originated in the asteroid belt, if only because the relative proximity of such a huge reservoir of free-floating debris makes it the statistically likeliest source. Analyzing the hot rocks when they do land, geologists have found that they fall into one of two chemical categories. The first group is far and away the more primitive one, made of some of the most ancient cosmic stuff imaginable. Superficially composed of iron and silicates, the samples also contain microscopic diamonds and tiny beads of a simple, primitive species of carbon. Such a composition is consistent not with the commonly accepted chemistry of planets, but with the commonly accepted chemistry of stars, and suggests that the meteors were cooked up long ago in some primal stellar oven—almost certainly the solar system's own sun. Indeed, so rich are the rocks in carbon and other primitive materials that they give off vapors of the stuff, tiny breaths of highly volatile compounds that even after billions of years still evaporate from the meteorite in pungent wisps. Astronomers with such an ancient sample in hand often amuse one another by dropping the rock into a jar, screwing on the lid, and coming back a few hours later to take a whiff of what's inside. The vaguely solvent-like scent, they remind one another soberly, is the oldest smell in the universe.

Ancient as such artifacts are, they are surprisingly easy to destroy. One way to annihilate them is to cook them down into a planet. As the gas, dust, and rubble that make up a fetal world begin to accrete, the gravitational power of the growing mass dissipates as heat, causing the world as a whole to reduce itself to a soft, hot, lava-like ball, much like early Earth and its early moon.

When bodies melt down this way, they undergo a dramatic transformation. The fragrant volatiles that so entertain contemporary scientists immediately boil away, unable to tolerate temperatures of even a few hundred degrees, never mind thousands. More important, the anatomy of the world itself begins to change, separating into layers as the heavier metals in the rocks precipitate toward the center of the body and the lighter ones remain on top. If a mass like this somehow fragmented back down into asteroids, each individual fragment would no longer be made of the mix of fundamental materials that characterize older, unmelted samples, but rather, of a single heavy element like nickel or iron, depending upon where in the planet's interior the rock originated. The second type of meteorites geologists recover generally have this type of homogenous composition. Since both types of meteorites routinely fall to Earth, it is likely that both types populate the asteroid belt. And if the rocks in the belt are made of this kind of material, rocks that wander out and become moons, like Metis, Adrastea, Amalthea, and Thebe, should, too.

That was the theory, anyway. But while the *Voyagers* could photograph the four inner satellites, the small size of all of the bodies and the fleeting look the spacecraft got at them made it impossible for the ships' chemical and light sensors to determine their makeup conclusively. The scientists could thus make all the inferences about the innards of Metis, Adrastea, Amalthea, and Thebe they wanted, but for now at least, they were powerless to confirm them. Mysterious as the guts of the inner four moons remained, however, the *Voyager* cameras were able to learn a lot about their surfaces.

Sixty-mile-wide Thebe, the highest flying of the four inner satellites, appears to have a lot in common with Earth's own 2,100-mile moon, at least when it comes to its history of meteor bombardment. One side of Thebe is scarred by three or four large impact craters, all nearly big enough to have fractured the moon entirely. Worse for the fragile satellite, the side that suffered the bombardment is the leading side—the one that faces forward as the moon moves through its orbit—meaning that Thebe took all of its biggest blows head-on. Had Thebe been able to rotate freely, the wounds it suffered would have been distributed more evenly across its surface. Like Earth's moon, however, the little Jovian satellite keeps one of its hemispheres pointed toward Jupiter at all times, locked in an orbital grip from which it can never turn away. Indeed, so tight is the gravitational hold Jupiter has on Thebe that the side of the moon that faces the planet actually bulges slightly. This small protrusion causes the mass on that side to increase, which in turn causes Jupiter's gravity to hold it even tighter, making it even harder for the moon to turn its face.

Metis and Adrastea, much smaller than Thebe, do not appear to have suffered the same kind of bombardment their larger sibling has, their modest size evidently allowing them simply to avoid many of the blows they otherwise would have absorbed. But while the two moons are not so remarkable for how they look, they *are* remarkable for how they behave. Lying even closer to Jupiter than Thebe does, Metis and Adrastea are deeper inside the planet's Roche limit, meaning that by rights they ought not to be there at all, having long since been crushed down into ring rubble. The fact that the moons do exist suggested to the *Voyager* scientists that the Roche limit rule has its exceptions and that its powerful gravitational hammer can be survived, provided that a moon orbiting a planet is small enough and sturdy enough. In the same way a large, creaking, multi-floor home will collapse in the face of an earthquake while a small stone cottage may emerge undamaged, so too will a relatively large moon—with all its flexing and churning and structural straining—suffer worse Roche damage than a static, asteroid-like rock.

But the mere act of surviving the worst the Roche limit has to dish out does not mean Metis and Adrastea are entirely unaffected by it. The moons' positions so close to Jupiter place them in direct proximity to the planet's rings—the presumed remains of a moon that hasn't fared so well gravitationally—and that has a curious effect of its own. Given the dusty nature of ring particles, any bands they form should be sloppy things, swirling at their centers and disintegrating at their edges before dissipating into space altogether. The bands around Jupiter, however, appear to be surprisingly sharp—tidy circles that maintain a relatively constant thickness across their entire width until they reach their outer rims, where they stop as cleanly as if they had been cut away by a blade. It is just beyond those rims that Metis and Adrastea are found.

Clearly, it appears, the two small moons act like a pair of orbital shepherds, moving constantly at the perimeter of the rings and rounding up stray particles that manage to escape. If Metis and Adrastea weren't there, the rings would probably still exist, but they would almost certainly be much messier, much more tenuous things, far less lovely for the loss of their two tiny attendants.

The last of the innermost moons, Amalthea, appears to be a far more passive player in the Jovian system, one that has not done much to affect the bodies around it, but has clearly been affected by them. Lying between Adrastea and Thebe, Amalthea is smaller than the Earthly island of Sicily and just as irregularly shaped, yet includes topographic formations seen nowhere on the distant Earth. *Voyager* scientists studying images of Amalthea spotted a sixty-two-mile-wide, five-mile-deep crater, which they dubbed Pan, and a second crater that wasn't quite as wide, but plunged twice as deep. Nearby stand two looming mountains, Mons Lyctas and Mons Ida, that appear to have been dug out and thrown up by the meteorite hits that created the vast pits. Soaring 12.4 miles above the crater basins, the mountains are more than twice the height of Mount Everest, a fact that is all the more impressive since the moon itself is more than sixty-eight times smaller than Earth. On a body this small, a mountain this big doesn't serve merely as topographic decoration, but as a sort of gravitational keel, helping to position the moon so that, as with Thebe, one face of the jagged world remains forever turned toward Jupiter.

More striking than the size of Mons Lyctas and Mons Ida is the color. Portions of the mountains' flanks—as well as the valleys and plains that lie below them—appear to be covered in an unmistakable dusting of ruddy red. Unlike Mars, the only other solid red body in the solar system, Amalthea doesn't appear to be uniformly colored. Rather, its surface is blotchy and uneven, almost as if the red did not originate on the moon, but instead was deposited from somewhere else. The *Voyager* astronomers believed they knew where that somewhere else was.

Barely 149,000 miles above the orbit of little Amalthea is the orbit of giant Io. On the scale of the Jovian system, 149,000 miles is just around the block, and for Amalthea that kind of proximity has real meaning. Linda Morabito and the other JPL scientists discovered a wealth of active volcanism on Io, and with the moon's low gravity, the red sulfur and silicate smoke emitted by the massive vents in the surface was unlikely to settle back down onto the moon. Instead, the emissions would simply be blown into Io's sky and permanently expelled into space. Chugging along downwind of Io, Amalthea would be unable to avoid the bigger moon's exhaust and—like a tugboat following a giant steamer—would soon be covered with the stuff.

"Io's volcanoes drive material up to enormous altitudes," said Torrence Johnson, another JPL planetary scientist. "When you take away atmosphere, the vent acts like a rocket nozzle, blasting the material much higher than it can go on Earth, and allowing it to settle on anything it might find in space."

The relationship between Io and Amalthea was far and away the most dramatic discovery the *Voyager* scientists made about the four inner Jovian satellites. The fact that they were able to learn even this much about moons as small and photography-defying as Amalthea, Thebe, Metis, and Adrastea surprised even them. When the *Voyagers* peered farther out into the Jovian system, toward the spots where the tiny, far more remote moons Leda, Himalia, Lysithea, Elara, Ananke, Carme, Pasiphae, and Sinope orbit, the scientific yield was, predictably, far less. Such distant, pebbly moons were completely beyond the reach of the *Voyagers'* sensing instruments, and even the spacecraft's cameras, built for just this kind of long-range snooping, were unable to return images that looked like much more than cosmic flyspecks. Nonetheless, the position and grouping of the specks was clear, with the eight moons clustered together in their character-istic flocks of four, and that at least told the scientists something.

When rubble gathers together so high above a world as these moons do, it probably did not simply accrete there, congealing out of the same cosmic raw material that formed the planets and the big moons. Rather, the rocks are likely the remains of deep-space aster-oids, bodies that flew in from the cosmos beyond Jupiter, got nabbed by the planet's gravity, and broke into pieces as they slowed down and dropped into orbit. The highly eccentric paths of all of the moons in the two clusters seemed to confirm this caught-on-the-fly scenario: Leda, Himalia, Lysithea, and Elara circle the planet at a 28-degree tilt, an orbital angle that is among the sharpest in the solar system. Ananke, Carme, Pasiphae, and Sinope, with their 105-degree inclination and their backward orbit, are more dizzying still. The composition of all eight satellites is probably similar to that of the more primitive, star-baked asteroids that fall periodically to Earth since the moons would have originated in the distant reaches of the solar system and thus would not likely be forged inside a planet that later fractured. Once again, however, given the tiny size of the worlds and the distance at which the spacecraft flew by, the JPL team could speculate about these matters, but *only* speculate.

The *Voyager* spacecraft's scientific haul was much greater when the ships turned their cameras—and the investigators turned their at-tention—to the four large Galilean moons. Of all the data Earth-bound scientists wait for when a spacecraft is flying by a large extraterrestrial body, there is perhaps none they wait for more anx-iously than a reading of the body's density. Density is the first impor-tant indicator of the world's composition, which itself helps determine chemistry, and from chemistry, everything else flows. Un-like radiation or temperature or magnetism readings—all of which can be read only with sophisticated on-board scanners and sniffers—density measurements can be taken with little more than the space-craft's own navigation system. Fly by a world with a gravity field powerful enough to pull hard on your ship, and you can be pretty sure you've flown by a dense world, one rich in metals and rocks and other heavy elements. Fly by a world that barely flicks your ship, and you've flown by one made of far fluffier stuff. Analyze those gravitational readings carefully enough, add a few magnetometer and spectrometer readings, and you can infer the anatomy of a large planet or moon as accurately as if you were down on its surface drilling.

When the *Voyagers* flew by the outermost of the four large moons Galileo had discovered—Callisto—the scientists were surprised to find that the ships seemed relatively unperturbed by the encounters. Though the moon measures 2,976 miles across, about the same size as the planet Mercury, it appeared to have only a third of Mercury's mass. While Mercury itself had long been known to be a sort of cosmic ball bearing—made up largely of iron—Callisto appeared to be only 60 percent metal and rock and the rest ordinary water ice. Moreover, unlike Earth and its own moon, whose interiors are stratified into mantles, crusts, and cores, Callisto appears to be a relatively homogenous place, with its rock and ice stirred into a well-blended mass that changes little until the very center of the world, where a small, rocky core might or might not lie buried.

For a moon with so lightweight a makeup, Callisto has survived a lot over the eons. The *Voyager* cameras revealed almost no patch of the Callistan surface that wasn't covered by craters. There were sea-sized craters and divot-sized craters, craters overlapping craters and craters within craters. There were even craters that formed in long chains—one stretching nearly four hundred miles—apparently created when an incoming asteroid was ripped apart by the gravity of nearby Jupiter, causing it to strike the surface of Callisto in a sharply angled, machine gun–like tattoo of small projectiles. A crust in which so many scars have been so perfectly preserved is a decidedly ancient surface, and the *Voyager* scientists concluded that the Callistan skin probably dates back to the creation of the very solar system, 4.5 billion years ago.

But while Callisto's craters are old, they are apparently not eternal. As the *Voyagers* dipped low over the moon, the cameras noticed some curious features. Unlike the largest craters on Earth's moon, which are surrounded by deep basin rings, the rings that surround the largest Callistan craters appear to be shallow, almost undetectable. While large craters on Earth's moon project bright rays from their centers—some of which can be easily seen from Earth itself—the Callistan rays are almost invisibly faint. Most dramatically, while the biggest basins on Earth's moon are circled by mountains driven up by the force of the incoming meteorite, there is no such uplift surrounding the basins on Callisto; the craters were dug out of flat plains, and plains the surrounding area remained, no matter how much surface material was thrown about by the impact.

To the geologists, the answer was obvious. The only sure way to preserve craters is to carve them out of something hard, like metal or rock. If you carve them, instead, out of something as ductile as ice, it won't be long before the surface begins to slip and slump, slowly softening the contours of even the deepest impact scars and erasing their details entirely. On a moon like Callisto, which was more than four-tenths ice, this surface relaxation would be going on constantly, and as it did, the face of the entire globe would be steadily reshaped.

Notwithstanding such constant crustal creep, the prevailing impression the scientists got of Callisto was of a static, almost inert world. With the moon's low density and relatively modest mass, it was simply not able to generate enough gravitational heat to stir up its innards in any significant way. But what if such a wet, chemically rich world managed to get just a little bigger and gravitationally hotter? What would happen if it could roil and warm itself in some meaningful way? The scientists found out when the *Voyagers* passed Ganymede.

Just what distinguishes a planet from a moon had never been scientifically certain before the *Voyager* spacecraft ventured to the outer solar system, but one obvious distinction was size. Planets, after all, could be mammoth things, with a world like Jupiter measuring more than 88,000 miles in diameter—or eleven times the size of Earth. Moons, by comparison, could be little more than cosmic pebbles, with some measuring just a handful of miles from one end to the other. As moons get larger, they become more and more planet-like, with a few—like Callisto or even Earth's own moon—pushing the very species line that separates a parent world from its satellite. Ganymede, with its 3,262-mile diameter, crossed that line entirely. If the gravitational cord that connects Ganymede to Jupiter were some-how cut and the moon were allowed to sail into orbit around the sun, it would not only readily join the family of planets, but also qualify as one of its more complex ones.

Though Ganymede, like Callisto, appears to be a relatively low-density rock-and-ice world, its sheer bulk is enough to give it a far more elaborate anatomy than its simple sister. Ganymede's original gravitational heat allowed it to stratify into no fewer than four dis-tinct layers: a small metallic inner core, a larger rocky outer core, a mantle of relatively warm ice, and a crust of harder, more brittle ice. A world with so differentiated a structure and a pure metal center can generate intense magnetism, and when the *Voyagers* switched their magnetometers on, they found that Ganymede—like Earth, but unlike Callisto—is indeed surrounded by a complex magnetic field.

The kind of subsurface fever that allowed Ganymede to develop so complicated an interior played a significant role in shaping the ex-terior of the globe, too. The surface of the moon appears to be a combination of two distinct topographies: dark, ancient stretches of meteorite-blasted land, and lighter, younger expanses of refreshed terrain, where extensive cracks and channels have allowed subter-ranean flows of some kind to well up and cover the worst of the world's impact craters. On a body as icy as Ganymede, it was a good bet that the upwellings weren't lava—not with the moon's rocky layer buried as deeply as it is beneath the icy crust and mantle. In-stead, the *Voyager* scientists were betting on water. As the once slushy mantle of Ganymede began to freeze earlier in the moon's life, it would behave like freezing water anywhere and expand. This would cause the overlying crust to stretch and fracture, allowing the slush to creep out and at least partially smooth the surface of the moon. The areas in which craters survived were those that had man-aged to hold back this warmer, upwelling flow. The areas that looked smoother had at one time been inundated.

If Ganymede was warm once, however, it clearly was no more. After its gravitational energy dissipated, the moon had only the heat of the sun to keep it from freezing over altogether. Half a billion miles from the center of the solar system, however, is a frightful place to hope for sunshine, and while Ganymede might have been forged in a furnace, it was now suspended in an icebox, its dynamic crust and mantle having long since gone still and cold.

The irony for poor, frozen Ganymede is that despite its cryogenic state, it is not located all that far from a source of world-warming energy. Just over 660,000 miles away lies the vast bulk of Jupiter, with its sizzling energy fields and its bone-crushing gravity. Just as Ganymede's bulk made it a slightly more dynamic world than light-weight Callisto, so, too, could a moon that lay closer to Jupiter be far more dynamic than Ganymede itself. Io and Europa fit that description.

When Linda Morabito spotted the volcanic plume streaming out of a crustal vent in the orange-yellow surface of Io, she became the first person ever to spot a volcano on the little Galilean moon, but not the first person ever to consider one. Only a few weeks before Morabito's discovery, a paper began circulating in the astronomical community theorizing that just such volcanic unrest might exist on Io. The evidence was not all that strong—nobody had actually seen volcanic exhaust yet—but it was nonetheless tantalizing. Both Earth-based telescopes and the *Pioneer* spacecraft had noticed a curious doughnut-shaped formation around Io, a cloud of particles that surround the moon in an all but invisible haze. There were only two possible sources for such a cosmic cloud: Jupiter, which could be sputtering off some kind of material that migrated into orbit around the moon; or Io itself, which could be outgassing volcanic exhaust that dispersed into nearby space. Most of the scientists who read the paper were betting on Io—and with good reason.

As the innermost of the four Galilean moons—orbiting just 261,000 miles above the planet, or roughly the same altitude at which Earth's own moon orbits—Io was about as deep inside the planet's gravity well as it was possible for a large moon to be and still remain intact. This kind of proximity has a profound effect on the satellite. While Earth's moon needs a full month to make one lazy circle around the planet's 25,000-mile circumference, Io is whipped around Jupiter's far larger, 279,000-mile equator in just 1.8 days. Io, like Earth's moon, is orbitally locked on Jupiter, with the gravitational hands of the planet keeping one face of the moon turned toward it at all times. In the Earth-moon system, this intimate gravity dance affects the very shape of the bodies themselves. The constant tug of the moon's gravity raises oceanic tides on Earth, causing the waters to recede and advance depending on which side of the planet is facing the always present moon at which point in the day. In return, the constant tug of Earth's gravity raises solid-body tides on the moon, temporary bulges in the lunar crust that subtly distort the overall shape of the globe. In the Jupiter-Io system, where there is a far greater disparity in the size of the bodies, things are all one way; Io has little gravitational effect on Jupiter, but Jupiter causes a solid-body tidal bulge on Io that rises more than three hundred feet high. What's more, since Io's orbit is slightly elliptical, the bulge tends to rise and fall, growing bigger when the moon is making its closest approach to the planet and smaller when it's at its farthest.

Such steady, wrenching changes in Io's shape would itself impose a lot of strain on the moon, but it is not Jupiter alone that has its gravitational way with Io. As the closest and speediest of the Galilean satellites, Io makes two complete orbits of Jupiter for every one its nearest sister, Europa, makes, and more than four for every one Ganymede makes. Each time Io passes one of these relative slowpoke moons, the neighboring world gives its orbit a quick gravitational pluck. As it does, Io strains briefly away from Jupiter, developing a tiny tidal bulge on its opposite hemisphere. The bulge quickly subsides, but no sooner does it vanish than another pass by another moon causes another one to rise. This much stretching and relaxing of the crust and mantle creates enormous subsurface friction, ultimately causing the moon's metabolism to speed up and its temperature to rise. As it does, volcanoes ought, in principle, to form.

That was the theory in any case before *Voyager* arrived at Jupiter, and when it did, Morabito discovered that the theory was explosively true. In the months *Voyagers 1* and *2* were barnstorming the world, volcanoes were discovered all over the face of Io. In just that time, whole new volcanic mountains rose up on the moon, whole new volcanic calderas opened, and whole stretches of Io's landscape were resurfaced practically overnight—by geological standards at least.

"The geology on Io changes on the scale of human lifetimes, on the scale of weeks or months," said Terrile. "If you lived on Io, you wouldn't have much need for weather reports, but you'd have an enormous need for geology reports: mountain building in the east, lava flows in the west. It happens that fast."

The eruptions that cause all these topographic changes are not only remarkably frequent but remarkably violent. The lava that the Io volcanoes emit gushes out at speeds of nearly 3,000 feet per second, ejecting more than 10,000 tons of material in a single blast. The moon's relatively thin atmosphere, which was found to include traces of oxygen and sodium, is rich in aerosolized sulfur, virtually all of it outgased by the explosions. Located so far from the incubator of the sun, the surface of most of Io rarely rises above −230 degrees; in the vicinity of the volcanoes, however, the temperature was measured at 60 degrees.

In addition to acting as a natural volcanic incubator, Io turned out to be a natural dynamo. Orbiting as close to Jupiter as it does, the moon lies deep within the planet's magnetic field. Flying so fast through such powerful lines of force, Io develops a current of nearly one trillion watts that crackle constantly across its diameter and leave a charged trail behind it as it soars through near-Jovian space. A moon that had once been thought to be little more than an inert sphere turned out to be not only thermally alive but also electrically alive, one of the most dynamic bodies anywhere in the solar system.

For all this violent energy, what Io lacked was some sign that it was organically alive, too. Unlike Ganymede and Callisto, this innermost of the Galilean moons is without any discernible concentrations of water, since whatever moisture it might once have had would have long since evaporated as a result of Io's tremendous internal heat. A world without water, of course, can never support known life. If it were possible to contrive a slightly different moon, however, one that could get itself even half as warm as Io and keep itself even half as wet as Ganymede and Callisto, that would be a moon to be biologically reckoned with. Such a moon, in the form of Europa, appeared to exist right next door.

When *Voyager 2* returned the first close-up images of the frozen crust of Europa, the JPL scientists knew they were looking at something remarkable. Located in the same punishing part of space where incoming projectiles had once done so much damage to the surfaces of Callisto and Ganymede, Europa should, by rights, have been covered with craters itself. *Voyager*'s first shimmery images of the moon, however, revealed nothing of the kind. Europa's crust, it appeared, was all but completely pristine; here and there might be a divot or two from a relatively recent meteorite hit, but apart from that, the battlefield look of the badly ravaged Ganymede and Callisto was utterly absent. What did mar the surface of Europa were cracks—hundreds, even thousands of them, forming a fine, spidery webwork across the face of the moon. It was the absence of craters that first caught the scientists' attention, but it was the presence of the cracks that held it fast. In both cases, they figured, they knew what was responsible.

As nearly as the *Voyager*s could tell, Europa is a relatively simple world. The smallest of the Galilean moons—just 1,945 miles across—it appears to be made up of almost nothing but core, a hard sphere of silicate rock nearly 1,800 miles wide. Surrounding this is a thin mantle of water ice and surrounding that is an even thinner water-ice crust. If the crust was uncratered, it must have periodically melted, and the *Voyager* scientists suspected that tidal pulsing was doing it. Though Ganymede and Callisto are too far from Jupiter to experience a significant amount of the kind of steady gravitational squeezing Io undergoes, and Io is so close it experiences too much, the roulette slot into which Europa fell as the moons were accreting around Jupiter was another matter. Located 415,000 miles from the planet, Europa was able to stay warm but not hot, gravitationally massaged but not gravitationally crushed. When an ice world is gently stirred up this way, it won't be long before it turns into a water world.

On Europa, that appeared to be precisely what was happening. As the moon's underground ice shifts and churns, it soon becomes slush, and the slush soon becomes water. The sugar-shell crust above it, in turn, continually thaws and refreezes, erasing any craters that once freckled it. And as the freshly refrozen crust grows stiffer and more brittle, it also repeatedly cracks, leading to the filigree of fractures that cover the world.

When the *Voyager* scientists peered into those cracks, they got another surprise. Most of the fractures were unexpectedly dark, suggesting both that water beneath the surface had welled up to fill them, and that that water was remarkably dirty. Given the prevailing chemistry of the Jovian system, JPL scientists were pretty certain that the contaminants that had muddied the water included salts, carbons, and various minerals, as well as nitrogen and dissolved oxygen and hydrogen, precisely the same organic materials that give lake water its murky appearance on Earth. Fill a warm, bathwater ocean with those kinds of ingredients, let constant tidal pumping keep the chemicals moving, and give everything 4.5 billion years or so to mix and marinate, and there was every reason to believe that life—ordinary, unremarkable, Earth-like life—would emerge.

"What we're finding with terrestrial organisms," said Terrile, "is that life has an extraordinarily high tolerance for conditions that we once thought were intolerable. There is almost no environment on Earth too punishing for some kind of biology to take hold as long as there's liquid water present. If there's water on Europa, too, then there are life forms on Earth right now with which we could probably contaminate that world's oceans. Whether there is any indigenous life on Europa already is the million-dollar question."

For the *Voyager* scientists, determining if a Europan ocean indeed exists beneath the ice, how warm and deep it is, and whether it is in fact home to its own native life-forms was, of course, pure conjecture. Answering those questions would require a closer look at the world, and a closer look at the world would take another spacecraft. For their part, the *Voyagers* were done with Jupiter. In the late summer of 1979, the second of the two ships flew beyond the last of the distant moons and vanished from the Jovian system forever, leaving the planet, its rings, and its flock of satellites far behind. Far ahead lay the other three giant worlds of the outer solar system—and separating those worlds lay hundreds of millions of miles the spacecraft would have to travel to reach them.

5

Cosmic Hoopskirt

Autumn 1980

When the engineers overseeing the *Voyager* mission arrived in the Space Flight Operations Facility late in the evening on November 12, 1980, they were not, by any measure, a relaxed group. Certainly, the walk over to the SFOF was pleasant enough. Getting to the big, boxy building on the west end of the JPL campus took you up the hill of Explorer Road, down the small path of Ranger Road, and through the slightly twisty stretch of Deep Space Network Lane, depositing you directly in front of the big, glassy entrance of the SFOF. The campus, as always, was largely empty by this time of night, and with the late autumn sun long gone, the streetlights on the footpaths had all blinked on, giving the grounds less the look of a large lab than a small village.

It was a pretty enough place to be all right, and on any other evening the *Voyager* engineers might have been able to enjoy it. On this particular evening, however, they had something else on their minds. What was troubling them mostly was the very real likelihood that the *Voyager 1* spacecraft—the spacecraft they were coming to the SFOF to fly tonight—was not going to live through the evening.

The engineers had gone through a lot to get to this night on which their spacecraft might well die, and the idea of losing the ship now was not an appealing one. It had been more than twenty months since *Voyager 1* had soared past Jupiter and its thirteen—actually, now sixteen—known moons, and the spacecraft had been in a flat-out sprint ever since. Clipping along at roughly 35,000 miles per hour, the ship had put more than 400 million additional miles on its odometer—or a tidy 20 million miles every month. Such a long-haul journey was just about sufficient to close the gap between the fifth planet from the sun and the sixth one, and before the evening was out, *Voyager 1* would indeed enter the neighborhood of Saturn, beginning its thirty-five-hour and four-minute trip past the planet, around its rings, and by at least seven of its eleven known moons, before flying out of the solar system altogether and vanishing into interstellar space. That was the plan anyhow, but standing between the spacecraft and this ambitious itinerary was the giant Saturnian moon Titan, and if the trajectory numbers streaming back to Pasadena were correct, the ship could easily be on a heading that would take it not past the vast moon, but directly into its rocky surface.

In the three-plus years *Voyager 1* had been in space, its flight plan had not changed significantly. As the ship flying point in this two-ship expedition, the spacecraft would be attempting the closest possible flyby of Titan, sweeping beneath its south pole, using the massive body's gravity to make a sharp turn north, and closely barnstorming the moon before it soared out of the solar system altogether. As long as *Voyager 1* succeeded, *Voyager 2* would be permitted to continue flying its horizontal course past Saturn and, possibly out to Uranus and Neptune. If *Voyager 1* failed, *Voyager 2* would be retargeted for the same acrobatic Titan flight.

In order to give *Voyager 1* any chance of pulling off the navigational stunt it would be attempting tonight, the flight dynamics engineers had charted a hair-raising route for it. According to the scientists' ciphering, if the ship was going to make the most of Titan's gravity field, it would have to fly just 4,023 miles above the bright orange cloud decks of the moon, burning its engine all the while to keep the ship's speed high and its direction true. Four thousand miles is a considerable distance by terrestrial standards, but terrestrial standards were not what was being applied this evening. When a spacecraft has already covered nearly 800 million miles on its looping trajectory from Florida to Saturn, a 4,000-mile miss distance does not amount to much. When the maneuver being attempted requires Earthbound engineers to transmit the spacecraft fly-by-wire signals that, even at the speed of light, take seventy-one minutes to reach the ship, the same 4,000-mile margin of error vanishes entirely. But 4,000 or so miles was all the *Voyager 1* engineers had to play with at the moment, and on this particular evening it would be Ray Heacock's job to see to it that that was enough.

Ray Heacock was, if anyone was still counting, the fifth project manager of the now nine-year-old *Voyager* program. First, there had been Bud Schurmeier, who oversaw things when the project was in its development phase; then came John Casani, who ran the show up through the launch; next was Robert Parks, who was in charge during the cruise to Jupiter; after that was Peter Lyman who took over from Parks for *Voyager 1*'s pass through the Jovian system. Now it was Heacock who had taken over the reins, and so far he had done a good job, commanding *Voyager 2* during its own Jupiter encounter and nursing both ships along as they made their transit to Saturn. At last, the first of those two spacecraft had arrived, and Heacock dearly did not want to become known as the only one of the five managers to wreck his ship on the shoals of his target world before he had a chance to collect so much as a single shred of data.

To hear Heacock tell it—when he cared to be completely honest, that is—there was no real reason he had to be in this position at all this evening, mostly because there was no real reason any spacecraft should be cozying up as close to Titan as this one was going to. It wasn't that Heacock didn't appreciate the appeal of the big, cloudy moon. Titan's sheer size, plus its rich hydrogen-oxygen-nitrogen chemistry, made it an irresistible target. The problem was, no matter how tempting Titan seemed, there were still dozens of other moons out there, too, and no matter how Heacock worked the arithmetic, he had never seen the sense in risking one spacecraft—and possibly two—studying just a single one of the solar system's satellites circling just a single one of its worlds.

Recently, a small but vocal group of dissenters at JPL had begun seeing things this way, too, and had begun arguing that perhaps it was time to consider removing a close flyby of Titan from the *Voyager* itinerary altogether. Heacock was part of this group and was not shy about making the case against Titan whenever the topic came up for discussion. The problem was, in the twenty months the ships had been traveling to Saturn, a larger, more vocal group of Titan advocates had taken precisely the opposite position, working hard to make sure the big orange moon stayed on the agenda. The most conspicuous of these pro-Titan spokesmen was astronomer and scientific pitchman Carl Sagan, and around JPL the accepted wisdom had always been that if you found yourself in a public debate over two competing scientific positions, Sagan was not the man you wanted at the opposite lectern. Equal parts Cornell University professor and television personality, Sagan had been serving as a consultant to the *Voyager* team for years. Whenever the conversation turned to Titan, he made it his business to champion the moon, touting it in that damnably seductive voice of his as "a uniquely interesting place," one where there was "undoubtedly organic chemistry going on." When Carl Sagan started going on about organic chemistry and the best you had to counter with was a lot of fretting and hand-wringing about 4,000-mile miss distances, there was little question whose argument was going to prevail.

The result was, on this November evening near the end of 1980, *Voyager 1* was indeed preparing to surf its way through near-Titanian space, and it would be up to a reluctant Ray Heacock to pull off the fancy flying that would make the encounter happen. For all his misgivings, Heacock knew he had at least a few things on his side—not the least being that his close approach to the moon would be over the friendly south pole.

Titan, as the *Voyager* team knew, flies a slightly inclined path around Saturn, one that takes it a bit above and a bit below the planet's equator—and thus a bit above and below its rings. Trying to fly a spacecraft across the Saturnian ring plane was not something the JPL flight planners had ever tried and not something they especially *cared* to try. While the rings were made up primarily of empty space, that empty space was still aswarm with tens of thousands of small and not so small particles. A mere grain of sand that collided with an object moving at the 35,000-mile-per-hour speed at which a *Voyager* traveled could pack the same energetic wallop as a bowling ball traveling at 120 miles per hour. A bigger bit of debris would hit harder still.

Happily for the scientists, Titan was, at the moment, orbiting slightly below the rings. This meant that as long as the spacecraft flew beneath the south pole of the moon before turning north, it would not even encounter the rings until it had already passed Titan completely. If there were any stray particles in the path of the *Voyager*, they would thus not have the opportunity to do any damage until after the spacecraft had done its work, taken its pictures, and beamed the data home to Earth. If the ship met an untimely end after that, it would be an end the scientists could accept.

Also boosting Heacock's confidence was that, hard as the kind of deep-space flying he'd be attempting tonight was, it was not completely without precedent. Just such a stunt had been pulled off nearly ten years earlier when William Pickering's *Mariner 9* was in orbit around Mars, and the JPL folklorists had been talking about the maneuver ever since.

The mission profile for *Mariner 9* had been an ambitious one, calling for the spacecraft to enter orbit around Mars—the first ship ever to do so—and spend eleven months circling the planet and sending home more than seven thousand pictures of its surface. Pickering's scientists had approved of this flight plan when it was first proposed, but secretly, most of them had wanted even more. As long as *Mariner 9* was out in Martian space anyway, they had wondered, why not up the exploratory ante and also send the spacecraft on a detour route that would take it by Phobos, the larger of the planet's twin moons? As tiny as Phobos is—just over thirteen miles wide—getting a good picture of it would require a breathtakingly close approach. That kind of maneuver was not something that had ever been contemplated, and for that reason as much as any other, the scientists had not initially raised the possibility of attempting it now. It was only when the ship was actually in Martian orbit and functioning perfectly that they drew themselves up and asked Pickering if it might be possible to give it a try.

Pickering, who was in no mood to risk the health of his spacecraft on an orbital thrill ride, was at first inclined to say no. The

more his scientists made their case, however, the more he realized it
was a good one. *Mariner 9* was there to study as much as it could
about the Martian system, and the moons were a critical part of it.
The only legitimate reason to resist the idea of steering his spacecraft
past Phobos would be if he didn't have confidence that his flight controllers could manage the maneuver; and if his flight controllers
couldn't manage such a maneuver, they shouldn't be flight controllers in the first place. To his own surprise as much as anyone
else's, Pickering okayed the plan, dispatching one of his scientists to
the SFOF to pass the order on to the men who would have to execute
it. The scientist went to the control center as instructed, tracked
down the chief navigation engineer, and begin explaining what he
needed.

"You know," he said, as nonchalantly as if the idea had just occurred to him, "we've been thinking upstairs that we'd really like to
get some close-up photos of Phobos."

"Okay," the engineer said absently.

"You can do that?"

"We can do that."

"How close can you go?"

"How close do you want to go?"

"As close as possible, I guess," the scientist said.

The engineer considered that for a moment. "How's twenty
miles?"

The scientist's eyes widened. "Twenty miles?" he said with a
slight laugh. "Sure, twenty miles is good." After a pause he added
warily: "Just don't hit it."

The engineer looked back at him fixedly. "We're not going to hit
it," he said and turned on his heel.

Before the day was out the men in the SFOF had figured out a
way to fly precisely the maneuver Pickering wanted. Before the week
was out the spacecraft had flawlessly flown it, sending the JPL scientists the best pictures ever taken of the little Martian moon. That, in
any event, was the way the story was told, and that was the way it
had *been* told since 1971. Now, on a cool night in 1980, Ray Heacock was going to have to live up to it.

When Heacock arrived in the SFOF to prepare for the Titan encounter, he quickly queried his controllers to be certain that both the ship and the men charged with caring for it were ready for the maneuver. When he had satisfied himself that they were, he instructed his flight dynamics officers to proceed, then listened from his console as they chatted among themselves, reading off the codes that would direct the ship to rotate to the appropriate attitude and fire its engine for the planned flyby. From there an engineer in a back room tap-tapped the command into a navigation computer, and from there the command was fired into space.

What happened after that was essentially nothing at all. Not only would it take the signal from Pasadena seventy-one minutes to reach the distant *Voyager* probe, it would also take another seventy-one minutes for the confirmation to come back that the maneuver had been executed. This 142-minute stretch far exceeded the length of time scheduled for the entire Titanian flyby, meaning that if the maneuver was a failure, the transmission *Voyager 1* sent back home would likely be a posthumous one, a beyond-the-grave shout that would not reach Earth until an hour after the spacecraft had already tumbled in toward Titan and annihilated itself on the surface of the moon.

Heacock spent the two hours and twenty-two minutes of waiting time uneasily pacing from his console in the SFOF to his office down the hall and back again, occasionally collaring a flight controller to ask him how he thought the maneuver was progressing. Each time the controller would answer with a no-problem nod that Heacock knew was either pure bravado or real confidence, and probably a little bit of both. Finally, as the 142nd minute of light-time concluded, Heacock resumed his seat, donned his headset, and waited, like the other men in the SFOF, for some word from space. After a few seconds a stream of numbers began to flicker onto the trajectory consoles in the middle of the room. The technicians watched as the data appeared, and after a moment one of them spoke evenly into his mouthpiece.

"A signal has been received," he said.

Heacock leaned forward, pressed his headset closer to his ears, and said nothing, waiting for the remainder of the call. Across the room most of his controllers mirrored him. A bare second or so later, the trajectory engineer called out again.

"Maneuver executed," he announced. "Titan encounter achieved."

Throughout the SFOF a cheer went up, applause broke out, and the controllers, who had been bent so worriedly over their screens, leaped up for a round of handshakes. Heacock, at his project manager's console, stood, too, and allowed himself a brief celebratory moment—but *only* a moment. *Voyager 1*'s near-Titan approach was not planned to be a long one. Arcing over the moon's south pole and up toward the north at a speed of 9.7 miles per second, the ship would pass Titan's equator in less than two and a half minutes and complete its reconnaissance of the entire moon in less than five. From there it would fly free in the Saturnian system, passing six of Saturn's other moons—Mimas, Enceladus, Tethys, Dione, Rhea, and Hyperion—as well as the giant ringed planet itself. In just over a day it would leave the Saturn neighborhood altogether.

This meant that the imaging team had to act fast. Even before *Voyager 1* began its Titan maneuver, the spacecraft had switched on its cameras and started beaming home pictures of the moon it was so closely barnstorming. Well before those signals had traveled the 800-million-plus miles back to Earth, the imaging team scientists had already gathered in their modest meeting room in the southeast corner of Building 264 to watch as the Titan portraits flickered to life on their multiple monitors. These first images, the scientists knew, would not be much—unenhanced black-and-white snapshots that would initially show only the moon's brightest features. Analysis of Titan's chemical spectra would allow the scientists to add inferred color to the image; subsequent refinement of the data stream would help them sharpen the picture and bring it into crisper relief. Nonetheless, for scientists trained to read interplantary pictures as a physician reads an x-ray, even the murkiest, milkiest Titan image could carry a wealth of information.

Just after the signal confirming the Titan flyby was received in Pasadena, word also came back that the first of the images had reached the deep-space antennas in the California desert, and was being relayed to JPL's massive mainframe computers, and from there to the viewing room monitors. For long minutes after the transmission arrived on Earth, those screens showed nothing but static. The scientists seated around the conference table stared fixedly at the glass, leaning expectantly forward when the static would flicker promisingly and then sitting back again when the flickering would stop.

Finally, after three or four such false alarms, the monitors began to blink and clear, and a faint ghost of a circle appeared. The circle faded, then reappeared, and at last sharpened itself into a stark gray-white ball. For a full five seconds, the scientists around the table stared at the image, and then, as one, they slumped.

The ghostly circle—the circle that was Saturn's Titan, the moon the scientists had waited more than three years to see—looked like nothing at all. There was not a shadow of a surface feature visible anywhere on it, not a hint of a cloud formation roiling its atmosphere. There was only the globe itself, an egg-smooth, egg-blank sphere, shrouded in an atmosphere that was completely opaque. For 325 years, astronomers on Earth had been peering at the moon from a distance of 800 million miles and wondering about its makeup. Now they were looking at it from an eyelash distance of just 4,000 miles and they were still left wondering.

"Nothing," somebody in the room muttered.

"A washout," somebody else responded.

None of the imaging scientists was prepared to say much more than that. There were plenty of pictures yet to stream in, and it was always possible that some of these would reveal something useful—a break in the clouds, perhaps, a storm in the atmosphere. Even without that, the scientists could still try to stretch and squeeze and brighten the data they did have, wringing at least a little more information out of all of the images. Intuitively, however, they knew that a little was all they were ever going to get. The moon that was perhaps the most promising world in the solar system save Earth itself was photographically blacked out by the very gaseous shroud that helped make it so chemically rich in the first place.

The scientists spent the better part of the next hour watching the blank, billiard-ball images of Titan arrive, and as each did, word seeped deeper and deeper into the JPL community that the grand visit to the grand moon had been an utter bust. Such a development should have hit the scientists hard, and indeed, the mood on the campus was grim. But there was cause to feel not so grim, too. First of all, there were still at least ten other Saturnian moons for *Voyager 1* to see, and before the day was out, images of those worlds should be in hand as well. More important, there was always *Voyager 2*. The impenetrability of the Titanian atmosphere meant that there had really been no reason for *Voyager 1* to visit the moon in the first place. And if there was no reason for *Voyager 1* to have gone to Titan, there was certainly no reason for *Voyager 2* to follow it there. The anti-Titan camp, it appeared, was going to carry the day after all. The second of JPL's magnificent deep-space ships was going to be able to remain safely in the plane of the solar system, where it could visit not just the Jovian system, not just the Saturnian system, but the Uranian and Neptunian systems that lay beyond.

Provided, of course, it was able to make it through its own Saturnian encounter alive.

August 26, 1981

Voyager 1's visit to Saturn was nowhere near the total loss it had started out to be. Just over sixteen hours after its fruitless flyby of Titan, the spacecraft passed within 258,000 miles of the midsized moon Tethys. An hour and twenty-nine minutes later, it came within 77,000 miles of Saturn itself. An hour and fifty-seven minutes after that it came within 55,000 miles of Mimas. That was followed by a 125,000-mile flight over Enceladus; a 100,000-mile approach to Dione; a close, 46,000-mile pass by Rhea; and, finally, a distant, 546,000-mile glimpse of Hyperion before the spacecraft vanished into space forever.

Nine months later, in August 1981, *Voyager 2* at last reached Saturn as well. Flying horizontally through the system, it first passed Titan at a great and indifferent distance of 413,000 miles. Next it flew by Dione at 311,000 miles, then Mimas at 192,000 miles, then Saturn itself at just 62,500 miles. Finally, it buzzed both Enceladus and Tethys at a distance of barely 54,000 miles and 58,000 miles, respectively, before sailing out of the Saturnian neighborhood just twenty-one hours and five minutes after it arrived, taking a last parting look at Rhea at a distance of 400,000 miles as it went.

Of all of the moons the ships saw after the disappointing Titan encounter, it was Enceladus and Tethys that most intrigued the JPL scientists. Measuring just 310 miles across, Enceladus is the second innermost of Saturn's moons and, as such, was thought to be gravitationally pumped by outlying Tethys and Dione, much the way Jupiter's Io and Europa are squeezed by the other Galilean satellites. The juxtaposition of the three moons alone suggested that this kind of gravitational dynamic might be under way, and the appearance of Enceladus—at least from the distant look *Voyager 1* got of it— seemed to confirm it. The moon seemed to be covered with a snowy surface material so bright it reflects nearly 100 percent of the sunlight that strikes it. Back on Earth, a full moon on a bright night reflects only 12 percent. This kind of ground cover would probably not lie exposed on the surface of Enceladus indefinitely without growing damaged and dimmer. The fact that it was as bright as it was suggested that some kind of tidal or volcanic force was causing upwellings or eruptions that were regularly refreshing the crust.

Neighboring Tethys was significant for just the opposite reason: the ancient, tortured appearance of its surface. *Voyager 1* discovered that Tethys, measuring 657 miles across, is scarred by a large crater and an associated crack that stretch nearly three-quarters of the way around its middle. The size of the scar suggested to scientists that Tethys was wounded almost mortally by the projectile that struck it. A slightly harder hit by a slightly bigger rock and the moon would have split open like a soft melon. Studying Tethys might provide clues to the internal stability of rocky worlds and how they are born and die. But while *Voyager 1* was able to confirm that the crack in the moon existed, the ship's quarter-million-mile flyby prevented it from determining much more.

For these reasons, getting good pictures of both moons was a

Voyager 2 mission priority, and in the additional months it took the ship to reach Saturn, the trajectory engineers worked hard writing the lines of programming code that would allow it to barnstorm the worlds. It was at the twenty-hour mark in *Voyager 2's* twenty-one-hour visit to Saturn that it was scheduled to make those Enceladus and Tethys flybys, and this time it was Esker Davis who was in charge. Davis was the sixth *Voyager* project manager, taking over from Ray Heacock in early 1981. Unlike Heacock, Davis would not have to worry about sending his ship through any acrobatic flips as it sailed through the Saturnian system. Instead, he would simply have to point the bow of his spacecraft straight ahead, steer the ship this way and that as he executed the navigational programs his engineers had developed for him, and prepare to collect the album of images the ship would send home.

Late in the day on August 26, the first of those pictures—from Titan, Dione, and Mimas—began to stream in, and by any measure they were dandies: clear, colorful, full of detail, everything the inscrutable Titan pictures had failed to be. So pleased was Davis with these early results that he promptly released a few of the preliminary images to the media. So pleased were the reporters with what they saw that they asked Davis to hold a late night briefing to describe what the pictures revealed so far. Davis readily agreed, summoning the reporters to a large conference room in Building 264 just before midnight. Holding up picture after picture, he pointed out some of the salient features in this one or that, begged the writers' pardon for not having had a chance to analyze them better, and promised to call another press conference after the sun came up, when there would be more science to report and more pictures to see, principally from the much-anticipated Enceladus and Tethys flybys. As he spoke, the reporters could listen to a public address speaker mounted on the wall behind him, crackling with round-the-clock chatter from the SFOF, where the laconic exchanges of the flight controllers provided the best possible evidence that the mission was proceeding well.

After just a few minutes Davis finished his presentation and the reporters left happy, making it a point to congratulate the new project director on his accomplishment so far and shake his hand on the way out. As the last of them was leaving, Davis, now alone, took it upon himself to shut off the lights in the conference room and close and lock the door. Just as he was doing that, however, he paused for a moment to listen to the SFOF speaker and heard something disturbing.

"Ace, this is Bus," an imaging technician called out to the flight director. "We have a problem."

Davis, who had spent plenty of hours himself in the SFOF, knew that there were big problems and small problems in any mission control room and that most of the time it was not the words of a flight controller that distinguished one from the other but the tone. The tone this time told him there was something seriously wrong, indeed. Slamming the door to the conference room, he ran across the campus to the SFOF, and as soon as he arrived saw what the problem was. The first pictures of Tethys had come in as scheduled, and they were just as crisp and rich as all of the others that had arrived that day had been. Good as the pictures were, however, they were also badly off center. In the first image Davis looked at, the moon was flying out of the right side of the frame; in the next one, it was flying off to the left; in the ones that followed, it was soaring out of the top or barely peeking in from the bottom. Davis quizzed the SFOF technicians and learned, to his distress, that the problem was not originating on the ground. A glitch in the guidance computers or imaging hardware at JPL itself could be fixed with little more than a kick. But the ground-based systems were working fine. The breakdown, it was evident, had almost certainly occurred in the body of *Voyager 2*, and from what Davis could tell, it was the ship's sensitive scan platform that was to blame.

In order to give the *Voyagers* the widest field of vision possible, the cameras on its instrument boom were mounted on a motorized platform that spun left and right, allowing the ship to aim its eyes without having to power up its thrusters and shift its entire body. Useful as an assembly like this was, it could also be maddeningly temperamental. Early in the *Voyager* missions, flight engineers had sent *Voyager 1* a signal instructing it to flex its scan platform for a routine check. The platform started to comply and then came to a stop with a grind the engineers could practically hear. The signal went out again, and the platform balked again. With the spacecraft utterly beyond any hands-on repairs, engineers hoping to fix it had to rely on the next best thing: a full-sized, working model of a *Voyager* that was kept in a clean room on the JPL campus. Put this understudy spacecraft through the same paces the real one had just gone through, and you could often re-create any breakdown that had occurred in space and determine what was behind it. In this case, the engineers re-created the problem easily and quickly found its probable cause.

Most of the mechanical assemblies on the *Voyager*s were insulated by multiple blankets of Teflon, a material that did a good job of protecting the spacecraft's instruments, but also had an occasional tendency to flake. In most cases, flaking Teflon was not a problem since any stray bits would simply drift harmlessly into space. When Teflon flaked near the scan platform, however, there was always the chance that a few bits could drift up into the motor assembly and cause the entire apparatus to seize up. That was what appeared to have happened to *Voyager 1*, and if that was indeed the case, the solution ought to be wonderfully simple: Just send the ship a series of commands ordering it to keep trying to turn, no matter what, and eventually the teeth of the gears in which the Teflon was caught should chew the contaminating flakes up completely. The SFOF controllers sent up the instruction, the platform did what it was told, and the Teflon, as predicted, crumbled away. Now, years later, *Voyager 2*'s scan platform seemed to have gotten itself hung up the same way.

Before concluding that that was indeed the problem, however, Davis wanted to be sure. Shortly before 1 A.M. he called his engineers back into the conference room in which he had just held his celebratory press conference and instructed them to bring him all of the performance readouts they could from both *Voyager 2*'s scan platform and the platform on the stand-in model in the clean room. The engineers complied and Davis immediately began comparing the data from both—placing the printouts side by side and running his finger down columns of corresponding numbers and along the peaks and troughs of corresponding graphs. As he did, he concluded that the profiles of the two just weren't similar enough—the numbers simply didn't line up as they ought to, the graphs simply didn't match. The way Davis read the readouts, the cause of *Voyager 2*'s off-center images was not Teflon at all; instead, it appeared to be an overheated actuator, a drive-train shaft that had lost its lubrication and was becoming jammed as it attempted to turn in its sleeve. Try to force the platform on *Voyager 2* the way the engineers had on *Voyager 1,* and he would likely rip the innards right out of the system.

The answer, Davis decided, would be to do almost nothing at all. Instead of forcing the platform, he would simply instruct his engineers to nudge it a bit, coaxing it slowly this way and that over the course of the upcoming days and weeks. In time, changing temperatures and migrating lubrication ought to help things loosen up a little, allowing the platform to spin free again. Davis announced this decision as soon as he reached it, and only hours after the crisis began the relieved project manager was able to send most of his engineers home for a little sleep. Returning to his office, Davis was preparing to collect his belongings and punch out for the day himself when his door flew open and a scientist from the imaging team burst in.

"What do you mean you refuse to command this platform?" the scientist half shouted.

Davis jumped a little and then forced himself to answer levelly.

"I'm not going to send any commands that are going to rip out the gears," he said.

"But we need Saturn pictures!"

"We've *got* Saturn pictures—20,000 from *Voyager 1* and 20,000 from *Voyager 2.* We probably have enough."

"But the press is still out there. We've got to send those commands."

Davis nodded no.

"Move that platform!" the scientists barked.

Davis nodded no. "We're not going to do that," he said to the obviously fatigued man. "We're just not."

"We'll see," the imaging scientist said. "I'll call headquarters." Turning on his heels, he stalked out of the office and slammed the door loudly behind him.

Davis stared at the door as it shook in its frame, rubbed his eyes, and stifled a yawn, idly trying to predict how long it would take before the scientist would regret his outburst. As it turned out, it didn't take long. Before the next day was out, the scientist—who by now had gotten a few hours' rest—reappeared in Davis's office with a mumbled apology and a promise that such a breach in protocol would never happen again. Davis told him to put it out of his head and dismissed the matter with a wave. Over the course of the next few weeks the engineers determined that *Voyager 2*'s problem was indeed not caused by Teflon at all, but by the very actuator problem Davis had suspected. Over the course of the months after that, the motor began to limber up. Well before it did, the JPL planetologists had already begun analyzing the pictures—both well-centered and off-center—that *Voyagers 1* and *2* had returned. Temperamental scan platforms or not, it was clear the ships had gotten what they came for.

●   ●   ●

Like centuries of astronomers who had observed Saturn before, the JPL scientists had a lot to look at when they studied the planet, but what claimed their attention first was its most conspicuous feature: its rings. The long-wondered-about ring system had never been terribly easy to study in much detail from Earth, but as nearly as astronomers could tell, it was made up of seven discrete bands, each of which appeared to be composed of a relatively homogenous river of rubble. The *Voyagers* found out that the structure is a good deal more complex than that. All of the bands in the Saturnian rings, it appeared, are made up of thousands of far finer strands, each just a fraction of the width of the bands as a whole and all of them nearly impossible to see from any appreciable distance. Up close, however, where the *Voyagers* flew, the ring system had the look of a grooved LP record, down to the distinct gaps that separate song from song or, in the case of Saturn, band from band.

Of the seven bands, the one closest to the planet is a precariously low-flying one, orbiting just 4,100 miles above the Saturnian cloud-tops. The band farthest away begins 74,200 miles up and reaches about 260,000 miles into space, or about the same as the distance between Earth and the moon. Despite their enormity, Saturn's rings, like Jupiter's, appeared to be made of extraordinarily tenuous stuff—ordinary clumps of water ice, a few of them with bits of rock at their center, but most of them as pristine as snow. While the largest of these frozen masses might be the size of big boulders, most of them are far smaller, measuring anywhere from a foot or two down to mere fractions of an inch. Viewed edge-on, Saturn's entire sparkly ring system is barely a mile thick, or sharper than a knife blade on its own massive scale.

Though the rings are almost certainly the remains of a lost Saturnian moon, it was clear that it hadn't been much of a moon. If the material that makes up all seven rings were scooped up, balled together, and tamped down into a single body, the re-created satellite would have a diameter of less than 120 miles. Such a modest body would have been easy to destroy with even a relatively small meteor hit, and most of the *Voyager* scientists agreed that that was probably just the disaster that struck the former satellite.

What impressed the scientists more than the fact that such a cosmic calamity befell the small moon was when it befell it. A ring formation this gauzy circling a planet this big would not be likely to have an especially long life span. No sooner would the moon be blasted apart and its remains dispersed around Saturn than at least some of the material would migrate inside the Roche limit. There, the planetary gravity would begin eroding the rings from the inside out, pulling more and more of the particles out of orbit and down to the surface, until the once broad bands crumbled to nothing. The fact that the rings exist today suggests that they are relatively new formations, existing for tens or hundreds of millions of years instead of the 4.5 billion years the solar system as a whole has been around. Had human beings emerged just a little earlier or a little later in celestial history, they would likely have seen no rings at all, coming to know a Saturn entirely denuded of the very feature that has so come to define it.

Short as the lives of the rings will be, they are lives that are being spent dynamically. Much of that dynamism, the *Voyager* team found, is caused by the planet's moons.

Not long before the *Voyager*s reached Saturn, astronomers working with ground-based telescopes were already adding to the planet's family of 11 known satellites, discovering tiny 19-mile Telesto and 16-mile Calypso orbiting 183,000 miles above the planet, and 20-mile-wide Helene orbiting 233,000 miles up. The *Voyager*s then spotted a cluster of three more moons: 19-mile Atlas, 62-mile Prometheus, and 56-mile Pandora, all circling the planet in close proximity to each other, between 85,000 and 88,000 miles up. All of the new moons were unlovely things—jagged, irregularly shaped objects that looked less like asteroids than moons at all. Nonetheless, all were undeniably part of the Saturnian family, raising the planet's total of satellite offspring to a whopping seventeen.

So much moon traffic moving through so much ring traffic made some kind of gravitational interplay inevitable, and the *Voyager* astronomers found plenty of it. Prometheus, like the rings themselves, was found to be a clean, porous body made principally of water ice, with a little studding of rock to give it some heft. Its small size and low gravity have helped it dodge most incoming meteors over the course of its long, long life, and Prometheus is thus relatively free of craters. As the little moon sails through its orbits, it appears to stay in close proximity to the inner edge of Saturn's narrow, 310-mile-wide F-ring—the fifth and one of the most tenuous of the planet's seven bands. Just on the opposite side of the F-ring, flying equally close to the outer edge of the fragile band, is Pandora, a body only slightly smaller and slightly more cratered than its sister. Plying orbital paths separated by just 1,457 miles, Prometheus and Pandora move at such similar speeds that they remain in near-lockstep, tightly straddling the F-ring as they move round and round the world. And just as Metis and Adrastea cooperate to keep part of Jupiter's ring formation tidy, so do Prometheus and Pandora continually groom Saturn's F-ring, keeping the zone of space on either side of it debris-free and the ring itself sharp-edged and neat.

A little closer to the planet, where the far broader, 9,000-mile-wide A-ring orbits, tiny Atlas performs similar sweep-up work. With less than half the size and gravity of Prometheus and Pandora—not to mention a much bigger ring mass to deal with—Atlas can do nowhere near the thorough job its two sister moons can. Nonetheless, it makes its gravitational presence felt. Patrolling the outer border of the sprawling A-ring, it manages to keep that edge of the band at least as clean as the F-ring's edges are—and certainly far cleaner than the A-ring's distant, unpoliced inner edge is.

Also doing its share of ring-cleaning work is the comparatively large, 243-mile-wide Mimas. Another relatively lightweight ice world, Mimas by rights should not exist at all. *Voyager* scientists found that this moon, like its sister Tethys, is scarred by an impact crater so big it nearly tore the moon apart. In the case of Mimas, the crater measures eighty miles across—or nearly a third of the diameter of the moon itself—and radiates stress cracks as far as its opposite hemisphere. The central peak of the crater—the cone of material that rebounds at the heart of an impact site—is nearly as tall as Mount Everest, climbing nearly four miles into Mimas's sky. On a world the size of Earth, this would be the equivalent of a mountain rising more than 131 miles above sea level.

Nonetheless, sturdy Mimas survived the hit and thus lived to help shape the appearance of the Saturnian system as a whole. Orbiting at the inner edge of the outermost ring—the vast, 186,000-mile-wide E-ring—Mimas seems to be associated with a huge, 5,000-mile-wide gap in the rings first spotted by Giovanni Cassini in 1675 and named, appropriately enough, the Cassini gap. The moon is certainly not large enough to have swept all of the material out of a stretch of space so vast, and some other body must have played at least some role in creating the Cassini gap. Mimas, however, did clear out much of it, helping not just to smooth the rings, but in this case to obliterate an entire stretch of them.

Slightly larger than Mimas—and infinitely more dynamic—is neighboring Enceladus. The bright, white surface of the 310-mile moon that *Voyager 1* photographed led astronomers to conclude that some kind of subsurface slurry was regularly resurfacing the world, and *Voyager 2* seemed to confirm that. Detailed images captured by the spacecraft as it buzzed the moon in its low, 54,000-mile approach revealed vast stretches of repaved terrain where new frozen flows appeared to have obliterated valleys, filled up basins, and in some cases cut craters cleanly in half, obscuring one side of them while turning the other into a sort of open-ended, topographic horseshoe.

Such slushy upwellings—cryovolcanism, as the *Voyager* scientists called the phenomenon—are probably due to the same tidal pumping that heats up Jupiter's Io, causing its more traditional silicate volcanism. While Io is squeezed by Jupiter itself and by nearby Europa and Ganymede, Enceladus is probably pumped by its sisters Tethys and Dione, both of which are more than double its size and orbit nearby, allowing them to pack a considerable gravitational wallop. So powerful is the constant pressure Tethys and Dione apply to Enceladus that the crystalline volcanoes that erupt across the smaller moon's face affect more than the moon itself. Orbiting 147,000 miles above Saturn, Enceladus is located deep inside the planet's huge E-ring. As *Voyager* scientists were studying the distribution of material in the ring, they noticed a curious thickening in the particle band that corresponds precisely to the orbital path Enceladus inscribes. The only sensible explanation was that the material comes directly *from* Enceladus, a sort of snowy, volcanic exhaust that the moon leaves in its wake as it sails around the planet, much the way a steamship trails a thick cloud of smokestack soot as it moves across an ocean.

"The peak of the E-ring in terms of density is right about the orbit of Enceladus," said Terrile, "so it's always been very suspicious. To all appearances, the moon doesn't merely shape the ring the way shepherd moons do. Instead, it actually *feeds* the E-ring."

To the *Voyager* scientists, it was no surprise that Saturn's satellites could play such a considerable role in determining the look of the planet's bands. The material that makes up the formations, after all, is the most fluffily insubstantial of stuff. Bring a massive, chugging moon into their vicinity and it would be hard-pressed *not* to distort things in some way. Far less expected than the impact the moons have on the bands is the impact they have on one another. Nowhere was this more evident than when the *Voyager* scientists turned their cameras toward Iapetus.

When Giovanni Cassini discovered the 905-mile-wide moon in 1671, he had no idea exactly what he was seeing. On the whole, a satellite circling a planet ought to be visible for very predictable portions of its orbit. When it is on the eastern side of its parent world, the moon ought to appear as a dot in telescopes as it moves counterclockwise away from the viewer. For a fixed period, the moon should then vanish as it passes behind the planet, eventually reappearing on the western side as it continues its counterclockwise path, this time moving toward the telescope. When Cassini studied Iapetus, however, he noticed something strange. No matter how many times he observed the moon, he was able to see it only when it was on the eastern side of Saturn, orbiting away from him. When it reappeared around the western side—or when it should have reappeared, at least—he saw nothing at all. Only if he was patient, waiting another thirty-five days for the moon to complete its half revolution and cross back over to the eastern side, was it once again visible.

Generations of astronomers who observed Iapetus after Cassini noticed the same phenomenon, though none of them could say with certainty what was causing the moon to flicker on and off this way. When the *Voyagers* flew by and looked up close, they provided the answer. Iapetus, the spacecraft's cameras discovered, is an almost perfectly two-toned world, with its forward hemisphere colored a deep, tarry black and its hindquarters colored a bright, snowy white. So dark is the prow half of the planet that it reflects as little as 3 percent of the light that strikes it—little brighter than asphalt. So bright is the rump half that it reflects a full 60 percent, or five times more than Earth's own moon. Since Iapetus is in a fixed orbit, keeping one hemisphere facing the planet at all times, its bright, stern end is always pointed toward Earth when it is on the east side of the planet, making it easily visible; its dark bow end is always exposed when the moon is on the west side, causing it, effectively, to vanish.

Without having a sample of the dark material on the front half of the moon, there was no way for the *Voyager* scientists to know precisely what it is made of, though its sooty color suggests that it is some kind of organic hydrocarbon. Where it comes from is more of a puzzle. Given the huge population of the Saturnian system and the gravitational turbulence that implies, it's always possible that Iapetus is tidally pumped by other worlds and simply releases the material from deep within itself. With no big moons within arm's reach of the world, however, such a volcanic process seems unlikely. That means the dusty stuff came from somewhere outside, and the *Voyager* team could guess the place.

Orbiting Saturn at a distance of more than two million miles, Iapetus is the second most remote of the planet's satellites. The one that earns the distinction of being the most remote is 136-mile-wide Phoebe, circling Saturn more than eight million miles high. Phoebe, unlike the other Saturnian moons, orbits the planet in a retrograde direction—clockwise instead of the customary counterclockwise. Also unlike the other moons, it does not orbit in the general vicinity of Saturn's equator, but rather flies at a sharply inclined angle of more than 175 degrees. For this reason, the *Voyager* astronomers concluded that Phoebe is not an indigenous satellite that accreted around Saturn, but rather an immigrant moon, a captured asteroid that was speeding through space until it became snared by Saturn's gravity.

If so, Phoebe, like other free-flying asteroids, ought to be made up of extremely primitive carbon compounds, and when the *Voyager* cameras caught a glimpse of the rocky world, they found that it is indeed colored the same telltale black as the asteroid bits that occasionally fall to Earth. This led the JPL scientists to concoct a complex scenario. If a wildly flying moon like Phoebe were to get whacked by another, smaller projectile—a collision that was by no means unlikely in the solar system's early, more violent days—a substantial chunk of its surface would get pulverized and blown into space. If this Phoebe-powder were then to spiral in through the Saturnian system—something that was also quite probable—it would settle on the first big object that got in its way, and that would likely be Iapetus, passing face first through the dark cloud the outer moon had given off. Every time Iapetus flew through the spot in space where Phoebe's debris hung, it would pick up more and more soot, growing darker and darker with each orbit. Ultimately, the leading half of the moon would have grown utterly black. From the point of view of Earthly observers, the spherical world would have been mysteriously reduced to a half sphere, one that both existed and didn't exist, depending upon its position at any given moment.

"For the most part the Saturnian system is clean as a whistle," said Torrence Johnson, "filled with moons made largely of ice. Then in the midst of it you have this half-blackened world, dirtied by processes that may take place on a different moon entirely."

Even more closely connected than the Phoebe-Iapetus pair is the Janus-Epimetheus pair. First spotted by Earthbound observers in 1966, Janus and Epimetheus, measuring 118 miles and 74 miles across, respectively, are two small, irregular moons orbiting about 94,000 miles high. When the trajectories of the two satellites were first analyzed, however, they were found to lie so close to one another that scientists studying them often grew confused, never quite sure which of the two moons they were looking at, or even if they were looking at two at all. The mystery stood for a decade and a half until the *Voyager* probes flew past the worlds and cleared up the confusion.

According to the images the spacecraft returned, Janus and Epimetheus typically orbit within thirty-one miles of one another, with one flying slightly lower—and thus moving slightly faster—than the other. Every four years whichever moon is on the inner, quicker track overtakes the one on the outer, slower one, causing the twin bodies to exchange a bit of momentum; this causes them suddenly to switch positions, with the lower moon boosting itself to the higher orbit and the higher one dropping down to the lower spot. Locked in this new clinch, they continue to waltz around the planet until four more years—or 2,100 more revolutions—have elapsed, when just enough new momentum will have built up to allow the moons to switch once again. The odd pirouette has probably gone on since the moons formed and will probably continue for as long as they survive.

A similar orbital fandango is danced by a trio of other Saturnian moons, Telesto, Calypso, and Tethys. Of the three, 657-mile Tethys—with its icy surface and globe-spanning crack—is far and away the largest and most complex, easily dwarfing the rocky, asteroid-sized Telesto and Calypso. Orbitally, however, the moons are co-equals. Flying at identical altitudes 182,689 miles above Saturn, the bodies form a neat, three-member procession, with Telesto walking point, Tethys flying behind it, and Calypso following them both as a sort of cosmic trace horse. The route the three moons follow around Saturn is reminiscent of the far longer march Jupiter makes around the sun accompanied by the two clusters of Trojan asteroids that precede and follow it. In both cases, the leading and trailing bodies are separated from the larger central body by precisely 60 degrees, the closest approach they can evidently make to one another without creating orbital instability.

Precisely 51,299 miles above Telesto, Calypso, and Tethys, another pair of moons, Helene and Dione, march in an orbital line of their own, also a precise 60 degrees apart. Helene, like Telesto and Calypso, is a mere jagged scrap of a world; Dione, like Tethys, is a bigger, rounder, icier one. Despite this disparity in size and gravitational power, the two moons are apparently able to maintain stable orbital positions, even without a third world like Calypso acting as a gravitational tailfin.

Of the remaining two moons in the Saturnian system, one of them—Rhea—held few surprises. Located one orbital slot above Dione and Helene, the 949-mile-wide moon turned out to have roughly the same snowball makeup as Dione, Tethys, and Iapetus. In its relatively isolated orbital perch, however, it is neither gravitationally linked to any other moons nor tidally pumped by any of them—meaning that it generally remains both cold and quiet.

The final remaining satellite, Hyperion, is another matter. One of the solar system's many irregular moons, Hyperion measures nearly 180 miles across. At that size and with that shape, the satellite is clearly pushing the limits of moon-forming physics. The larger a body gets, the likelier it becomes that the force of its accumulated gravity will cause it to collapse on itself, making its highest peaks crumble, its deepest valleys fill, and the moon as a whole assume at least an approximate spherical shape. Hyperion is almost big enough to round out this way—and indeed, it almost certainly would have if it had managed to accumulate just a bit more mass. Instead, it has remained forever imperfect, its slightly-too-small size keeping it permanently poised at the brink of the sphere it clearly wants to be, but never actually letting it make the transformation.

Hyperion's irregular shape is matched by its irregular rotation. Unlike most other satellites, which spin on their axes at a fixed and predictable rate, Hyperion is in a permanent state of rotational chaos, somersaulting through its orbits in a constant tumble that causes it to point toward a different spot in space at any given moment. At least part of the reason for this erratic motion is Hyperion's violent past. The little moon appears to be heavily cratered, with one impact scar stretching seventy-five miles across its face—nearly half the length of Hyperion itself. The moon's dense, rocky interior—far denser and far rockier than that of most other Saturnian moons—probably prevented it from being destroyed by this bombardment, but it could not prevent it from being knocked off its pins, utterly losing its bearings and never quite regaining them.

A moon this erratic, of course, is not the rule for Saturn. With the exceptions of Hyperion and high-flying Phoebe, things generally remain orderly in the Saturnian system, with clean snowball moons sailing through light, fluffy rings, all of them maintaining a sense of near-perfect gravitational order. Most of the seventeen moons have some of the essential ingredients of simple biology—carbon, water, perhaps some free oxygen and hydrogen—but without sufficient gravitational turbulence to keep them terribly hot, they are unlikely to make much of these raw materials. It was only Titan—with its rich, opaque haze of chemistry—that stood any chance of achieving something biologically extraordinary, and for now, at least, Titan would remain a mystery.

On August 26, 1981, about 4.5 billion years after Titan and its sister moons were formed and about four years after two tiny rocket ships left the ground on a peninsula of land on faraway Earth, *Voyager 2* left the Saturnian system behind, heading in the general direction of Uranus, the next world down the planetary line. It had been exactly two hundred years since William Herschel discovered the Uranian system, and in all that time it had eluded detailed study. In just four more years it might at last be seen up close.

6

The Edge

January 28, 1986, 8:30 A.M.
California Time

Most of the people at the Jet Propulsion Laboratory had heard all they cared to hear about the seven astronauts aboard the space shuttle *Challenger.* The press could write about them all they wanted, the public could read about them all they wanted, but the folks at JPL were just about fed up.

It had been almost five years since NASA's celebrated space shuttle had started flying—the maiden ship going up in April 1981, just four months before JPL's own *Voyager 2* probe made its historic flyby of Saturn. At the time the huge new space plane had been nothing short of a sensation, taking off like a rocket, landing like a jet, and shedding almost no expendable parts along the way. But sensations fade and years pass, and what was novel once becomes not so novel anymore. By January 28, 1986, NASA was ready to launch yet another shuttle—the twenty-fifth such mission, according to the space agency's reckoning—and by all rights nobody should really have cared.

The fact that so many people did care this time around had nothing to do with the work the ship was being sent up to do—though the workload on this flight was considerable. Rather, it had to do with whom NASA was sending up to do it.

Ever since the first shuttle flew, NASA had been boasting that the era of high-risk spacecraft had come to an end. The shuttle would be so safe and reliable a spacecraft—a space*liner,* really—that it would no longer even be necessary to send only pilots aboard. While at least two skilled stick-and-rudder men would always have to fill the shuttle's two main cockpit seats, the other five seats in the ship could now be occupied by almost anybody. Over the years the space agency had made good on this promise, filling out the shuttle's flight manifests with biologists, physicians, and even politicians, all civilians who never would have remotely qualified for a mission before. Now, after two dozen shuttle missions, NASA was set to do it again. Included in the five-man, two-woman crew of the space shuttle *Challenger*'s current mission was a previously unknown New England schoolteacher, a soon-to-be space traveler who had spent the last year training for a mission that once again would help the space agency prove that in the modern era it wasn't necessary to *be* an astronaut in order to fly like one. For this reason as much as any other, the upcoming *Challenger* flight had generated immense publicity, and for the scientists at JPL, who could always use a little publicity of their own, it had generated it at the worst possible time.

In 1981, long before the *Challenger* crew had been chosen—indeed, long before all but the earliest shuttle flights had even flown—NASA had made a quiet decision. Somewhere in space, a billion or so miles from Earth, the *Voyager 2* probe had just passed the planet Saturn and headed off in the direction of Uranus. By rights, *Voyager 2* had no business going anywhere near Uranus since NASA had made it clear from the moment the spacecraft left the ground that it was not budgeted for any mission beyond Saturn. But now, with the ship having completed its visit to the big ringed world, and with its systems operating so well you would almost think, if you didn't know better, that someone had designed them for more than their stipulated three years of life, it didn't seem scientifically sensible—or even scientifically *moral*—not to let it fly on and see where it could go. For years JPL engineers had been making the case for just this kind of extended mission, but the NASA administrators had always said no. In an agency as big as NASA, however, minds change and policies change, and by 1981, when the *Voyager* ships had already returned a dazzling portfolio of images from Jupiter and Saturn, it didn't take long for that no to change to a yes.

On January 24, 1986, just four days before *Challenger*'s intended launch, the *Voyager 2* spacecraft was scheduled to reward the agency's faith in its abilities, completing its fifty-three-month, 916-million-mile journey from Saturn to Uranus, flying directly through the heart of the densely populated Uranian system and out the other side. The pictures the ship sent home from this encounter promised to be dazzling, but in order to take them, the sometimes balky *Voyager 2* would have to do some fancy flying.

Long ago ground-based scientists had discovered that Uranus is a planet like none other in the solar system. While the other eight planets all orbit more or less upright, Uranus is tipped loopily on its side, completely toppled so that it rotates horizontally, with one pole pointing back toward the sun and the other pointing off into space. The planet probably assumed this drunken position as a result of a collision long ago with a projectile at least as big as Earth. Whatever tipped the world over, such an odd posture meant no end of navigational headaches for spacecraft trajectory planners. Stars flickering on and off in the vicinity of the planet indicated that Uranus probably had at least nine faint rings, all of which were also turned at a sharp angle to the horizontal. Also circling the planet were five comparatively large moons—Miranda, Ariel, Umbriel, Titania, and Oberon, ranging from 292 to 978 miles in diameter—that orbited more or less vertically as well. How many other smaller moons there might be, flying at how many other vertiginous angles, was impossible to say.

Tiptoeing through this planetary train wreck, and doing so in a way that would give *Voyager 2* any hope of flying on to Neptune afterwards, would require some creative navigation—and some necessary compromises. The way the mission overseers figured things, if they were going to get out of the Uranian system alive, they would be able to make a close flyby of only the smallest of the known Uranian moons—Miranda—using its comparatively gentle gravity field to nudge the ship along. This trajectory would help the ship keep a safe distance from both the planet itself and the other satellites—a prudent precaution, but one that would prevent it from taking anything better than arm's-length pictures. Nonetheless, arm's length was still breathtakingly closer than any human or robotic eye had ever gotten to Uranus before, and the JPL scientists were willing to settle for almost any look their spacecraft could give them.

Preparing for this high-wire encounter with Uranus had been a long process. In almost any deep-space mission, cruise time—the time it takes a ship to travel to its intended destination—could be dead time, with nothing for either the spacecraft or flight controllers to do but wait for the ship to get where it was going. Between Saturn and Uranus this four-year-plus period of idleness was longer and deader than most—so dead, in fact, that *Voyager 2* could essentially be put to sleep. Not long after the spacecraft passed beyond the harbor lights of the Saturnian system, its cameras were shut off and most of its instruments were powered down. Now and then a few sleepy sensors might be prodded to life to take this or that measurement of this or that particle field. But for the most part, the ship was kept electronically anesthetized, programmed only to send home a regular readout of its own vital signs and to sound an alert if one of those signs began to flicker or fail.

For this kind of graveyard-shift work, a team of trained scientists was not necessary, and at the same time *Voyager 2* was being unplugged, most of the people responsible for caring for the ship were reassigned to other missions. Shortly before the spacecraft at last reached Uranus, those on-loan scientists were recalled and put back to work preparing for the encounter. Once they were, however, they found to their distress that even as *Voyager 2*'s cameras were pointing toward the steadily growing blue-green world 1.7 billion miles away, the media's cameras were stubbornly pointed toward the launch pad at Cape Canaveral, where the *Challenger* crew was preparing for a decidedly less ambitious mission two hundred miles into space.

Nonetheless, as the January 24 encounter date approached, the members of the *Voyager* team worked hard to publicize the flyby to any member of the media who cared to pay attention to it. By now, Esker Davis had stepped aside as manager of the *Voyager* project and had been replaced by Richard Laeser, a JPL scientist who had worked on a number of planetary projects before and had come to appreciate that one of the first jobs of any project manager was not just to fly a mission successfully but to promote it successfully, selling both the science and the spacecraft to a public that didn't always want to buy. When it came to *Voyager,* he knew, the project managers had always promoted well.

As the first *Voyager* was approaching Jupiter in 1979, reporters from all over the country clamored to be on hand in Von Karman Auditorium for the encounter. While the crowd shaped up to be a good one, however, it was hardly standing room only. In order to fill things out and avoid the embarrassing spectacle of a half-empty Von Karman on national TV, the public affairs office decided to loosen its media standards a bit, granting credentials to reporters from science fiction magazines and other fringe publications who never would have earned a press pass before—the thinking being that a flaky journalist and a mainstream journalist would look pretty much the same to anyone watching the press conference at home on TV.

By the time the Saturn encounter arrived, things had changed dramatically. The Jupiter pictures *Voyagers 1* and *2* beamed home had created such a stir in the media that the requests for Von Karman press credentials soared. The public affairs office once again had the luxury of picking and choosing who would attend the event, and this time the science fiction writers would be excluded. When the Uranus flyby approached more than four years later, competing so futilely with the *Challenger* launch, the demand for access predictably dipped. Regardless, Laeser decided to make the most of the event, scheduling as big and ceremonial a press conference as all of the project managers who preceded him had, and indeed making his bigger still, flinging open the doors not only to Von Karman, but to a smaller press room adjacent to the main hall. Reporters of all stripes would be invited to attend, and Laeser, eschewing his customary seat in the control center, would be there, too, circulating through the two rooms, shaking the hands of his hard-core media supporters, and making sure they knew how grateful his lab was for their ongoing attention.

On the morning of the Uranus flyby, the turnout was a bit better than Laeser had expected—and the show the reporters got was more spectacular. *Voyager 2*'s initial encounter was its close approach to Miranda, a maneuver that would take it to within 19,000 miles of the moon. The rendezvous would occur early in the day, but the signals, traveling 1.7 billion miles from the Uranian system, would need over two and a half hours to arrive. After that they would need another few minutes to be assembled, and only after that would the crowd in both press rooms see the sights that *Voyager* had seen. When the spacecraft's pictures at last did appear on the JPL TV screens, those sights moved the Von Karman crowd to speechlessness.

Glowing on the multiple monitors in the press rooms was the familiar circle of an otherworldly world, filled with one of the most unfamiliar landscapes the reporters had ever seen. The surface of Miranda was a fractured, fragmented, jigsawed mess, a smashed-looking place in which a smooth plain lay hard up against a jagged scarp which itself lay hard up against a capsized cliff or a deep valley. It was almost as if Miranda had somehow been utterly shattered, broken into countless ragged bits, and then had reassembled itself into any pattern that best approximated a sphere.

For a long half minute after the images appeared, the normally voluble reporters simply stared; then, at last, a cheer went up. Laeser, standing in the midst of this, was quickly set upon—his hand shaken, his back slapped, his attention pulled this way and that as the reporters shouted a volley of questions. The scene was repeated again and again throughout the morning, as the speeding *Voyager 2* sent back snapshot after snapshot of the tortured Miranda. After that, the ship proceeded through the Uranian system, returning equally sharp images of other moons, twinkly rings, and at last, Uranus itself, before finally flying away from the planet and heading in the general direction of Neptune. For this one morning, in this one room, Laeser could convince himself that there was no *Challenger* poised on its launch pad in Florida, turning so many media heads as it readied itself for its launch later in the week. Today there was only *Voyager* and Uranus and the clutch of reporters who came to see what the ship and planet had to tell them.

Now, four days later, at 8:30 A.M. on the morning of January 28, a more sober public relations reality had returned. The *Voyager* scientists had had a good ninety-six hours to study the batch of pictures the spacecraft returned and were ready to discuss their preliminary findings with the press. Such a follow-up briefing, Laeser and the others knew, was where the science in a mission like this started getting done, where the geology and planetology work the spacecraft had been built and launched for in the first place was at last made public. This, in theory, ought to be what really interested the reporters, but today, even the comparatively modest crowd from earlier in the week would not be coming around. The press conference at which the scientists would make their first disclosures had been called for 10 A.M. Pasadena time. Only hours after that schedule had been set, word had gone out to the space community that *Challenger*, which had been waiting out bad weather in Florida, had at last been cleared to fly and would be taking off at 8:38 A.M. If the liftoff had been set for later in the day, there was at least a chance that the reporters would take in the JPL conference first—if only by closed-circuit TV—and then turn their attention to the *Challenger* flight. With the space shuttle leaving when it was, however, virtually the entire press corps would be busy for virtually the entire day.

Nonetheless, Laeser once again decided to hold his conference as scheduled and hope for the best turnout possible. While waiting for the event to begin, he and his senior staff figured they might as well do what the reporters all over the country were doing, and gathered in the fourth-floor conference room of Building 264 to watch the *Challenger* launch on TV. By the time the liftoff was just a few minutes away, most of the seats at the room's thirty-person table were filled with *Voyager* scientists, all fixing their attention on a cluster of color TVs, which at the moment were showing an image of the bright white *Challenger* sitting on its pad, sweating vapor into the blue-gray Florida morning.

The men followed the countdown until the engines ignited, the shuttle shuddered, and the entire rattling assembly of spacecraft, fuel tank, and solid boosters muscled themselves into the sky. They watched the ship fly for fifteen seconds, thirty seconds, then nearly a minute. They heard the commander, Dick Scobee, announce, "Houston, we have roll program," as the spacecraft turned on its long axis to orient itself for its sprint to space. They heard Richard Covey, the spacecraft communicator in Mission Control, radio back, "Roger, roll, *Challenger*." They heard the murmuring voice of the mission narrator confirming that the ship had three good fuel cells, three good engines, three good power-output units.

Finally, they heard Covey call out to Scobee, "*Challenger*, go with throttle up." They heard Scobee call back, "Roger, go with throttle up."

What they didn't hear a few seconds later—what nobody watching the launch on television heard—was the lone voice of a flight controller sitting at a console in Cape Canaveral who looked down at the monitor in front of him and saw, to his horror, that the dappling of numbers that usually filled the screen had suddenly vanished, the monitor going as blank as if the shuttle had suddenly come unplugged.

"Where in hell is the bird?" he called out to no one and everyone.

On the television screens in Building 264—and on millions of screens around the country—it became immediately clear where *Challenger* was, and where it was was nowhere at all. The *Voyager* scientists watched as the tiny image of the rising ship all at once transformed itself into a much bigger circle of smoke. The circle then swelled into a fireball and the fireball grew two ugly fingers.

"What *is* that?" someone in the room asked, knowing full well what it was.

"That isn't right," someone else said, knowing with certainty that it wasn't.

"Obviously a major malfunction," the voice from the television said self-evidently and absurdly.

Dick Laeser and the rest of the *Voyager* team watched slack-jawed as the death of the *Challenger* played out—as the ship blasted its vitals across the Florida sky, as those scorching remains fell sizzling into the ocean, as the crowd at Cape Canaveral pointed and gaped. When the drama was over, the Pasadena scientists stared stunned and red-eyed at one another, saying nothing because, after all, there was not a thing to say. Laeser, the Earth-based commander of a much smaller spacecraft that earlier in the week had had a much better day, then walked woozily back to his office, sat at his desk, and buried his head in his hands. After a moment he picked up the phone and called his director of public affairs.

"Cancel the press conference," he said hoarsely. "Just cancel it."

Nobody, least of all Laeser himself, cared a whit about Uranus this morning.

•　　•　　•

It would be several months before anyone in the larger space community thought about much beyond the loss of the *Challenger*. First there had to be the burials, then there had to be the wakes, then there had to be the inquests and the investigations and the hurled accusations. Then, finally, there was the inevitable resolution: the vow to build a better ship, a safer ship, one that would never die so gruesome a death again. With that promise, the scientists and reporters and public at large were at last ready to look skyward again, and when they looked, they found Uranus and its moons and the pictures *Voyager 2* had returned of both waiting for them. What they learned from the images was both more and less than what they'd hoped for.

Superficially, Uranus had promised to be something magnificent. With a diameter of 32,000 miles, it was the fourth largest body in the solar system, four times bigger than the modest Earth, and more than 10 times bigger than a nugget world like, say, Mercury. Uranus's vast, swampy atmosphere would almost certainly be a dynamic one. Its fields of radiated energy would almost certainly be crackling ones. It would, on the whole, be a planet well worth studying. When *Voyager 2* looked toward the planet, however, it essentially saw nothing at all.

Uranus, the JPL scientists discovered to their disappointment, was Titan all over again. Instead of Titan's characteristic orange, Uranus was a soft blue, but with that exception, it looked precisely like the far smaller Saturnian moon had looked. There was not a scrap of a cloud anywhere in its atmosphere, not a swirl or an eddy breaking up the blankness of its face. There was only a vast, undifferentiated gas ball that gave up no visual clues to its composition or its workings. Chemical sensors aboard the spacecraft *were* able to sniff out the makeup of the planet's atmosphere and found it to be nearly 83 percent hydrogen, 15 percent helium, and 2 percent methane, plus some other stray hydrocarbons. Other instruments were able to discern a curious, corkscrew magnetic trail following Uranus as the planet simultaneously rotated on its side and revolved around the sun. Apart from those barely illuminating findings, however, Uranus showed *Voyager* a face that was utterly expressionless.

But *Voyager* hadn't necessarily come to see the planet, it had come to see the objects that circled the planet, and those objects delivered on their promise. The ring system that Earth-based astronomers thought they saw through their telescopes indeed existed, and indeed orbited upright, circling the planet's sideways equator. The rings were fine and faint formations, but not so fine and faint that the cameras didn't reveal two additional ones that had never been spotted from Earth, bringing the total number of the fairy-dust bands to eleven. Like the rings around Jupiter and Saturn, the Uranian formations were probably the remains of a shattered moon, one that previously circled the planet intact but at some cataclysmic moment in its history was reduced to rubble. And like the other planets' rings, these probably would not last forever, decaying and crumbling and eventually falling out of orbit altogether.

The death of this one moon and the ring it became didn't mean that the planet couldn't be circled by a whole family of surviving moons—and a big family at that. As poorly photographed as Uranus had always been from Earth and as small as some moons could be, JPL astronomers had little doubt they'd discover many more satellites orbiting the planet than the five that had already been spotted by telescopes. In anticipation of this, Steve Synnott, who had hauled in so many moons when the *Voyager*s flew by Jupiter, had developed new satellite-hunting software that would make the job of spotting undiscovered bodies far easier than it had ever been before. Rather than assigning humans to the tedious squint-and-compare business of studying stacks upon stacks of pictures, trying to find a single moving dot in the vastness of Uranian space, Synnott instead auto-mated the job. An imaging team member who found a single suspi-cious point in one picture could load that image—along with hundreds of subsequent images—into a computer and instruct the machine to scan the pictures and alert the scientists if the point seemed to move. If it did, the scientists could then calculate the na-ture of that motion to determine if it looked like orbital motion. If the motion was predictable enough, they would know they had themselves a moon.

Nonetheless, it was still up to human beings to tell the computer which little pinpoints to study, and that took work. The images that poured back from the spacecraft were messy ones, filled with noise and static and electromagnetic freckles, all of which showed up on the JPL television screens, and any of which could be a Uranian satel-lite. What's more, the screens themselves were imperfect, with blown pixels and other blemishes occasionally creating distracting dots of their own. Over time, the JPL researchers learned to recognize the unique flaws in each of their monitors, and took to drawing tiny grease pencil circles at random spots on the glass screens, reminding themselves that any dot they found inside a circle was probably not a moon.

Even with all of those distractions, the Uranian moon harvest turned out to be huge. The five known and relatively large moons had all been tracked in orbits between 80,500 and 362,000 miles high. When *Voyager 2* pointed its cameras closer to Uranus and focused them down to spot smaller objects, a whole school of tinier moons swam into view. As nearly as the scientists could make out, there were at least ten previously unknown, boulder-like bodies circling the planet, measuring between 16 and 95 miles in diameter. All of the moons were dark, sooty things, reflecting barely 7 percent of the light that struck them. Such floating lumps of coal looked to be exceedingly old and exceedingly rich in carbon—one more piece of proof that the solar system as a whole is fairly awash in the fundamental stuff of bi-ology. The moons' reasonably orderly orbits suggested that they were not captured bodies snatched out of the skies by the Uranian gravity, but instead formed in place when the planet itself did.

Unremarkable as the moons looked, they were given some decidedly lyrical names. Synnott and the other JPL scientists largely adhered to Uranus's Shakespearean naming tradition, christening the first nine new moons Cordelia, Ophelia, Bianca, Cressida, Desdemona, Juliet, Portia, Rosalind, and Puck. The tenth one, Belinda, was named after a character in Alexander Pope's *Rape of the Lock*. Apart from Cordelia and Ophelia, which straddle and shepherd Uranus's bright Epsilon ring, all of the small moons were found to orbit essentially alone, having little gravitational influence on the planet, the other rings, or even one another. As heavenly bodies went, the ten new moons were thus little to speak of, significant more for their very decorativeness than for any new scientific insights they offered.

The larger moons were a different matter. As a rule, the planets of the inner solar system are denser than planets in the outer, mostly because the blowtorch force of the nearby sun blasts away most lighter gases, reducing Mercury, Venus, Mars, and even Earth to small, relatively rocky pellets. More distant worlds like Jupiter and Saturn, which feel far less of the sun's effects, retain their gases and thus tend to be far larger and more vaporous. When it comes to the moons that circle most planets, this density rule also applies, but for different reasons.

Out in the solar system's successively more remote regions, temperatures fall lower and lower, with the thermometer in the Neptunian system approaching −400 degrees. The four large planets, which generate their own gravitational and radioactive heat, are able to compensate for this, keeping their insides hot and preventing their atmospheres from freezing. The planets' moons have far weaker internal furnaces—and often have none at all. With no way to keep themselves warm, they are thus completely exposed to the deep freeze of deep space, growing steadily colder as they move farther away. At such paralyzing temperatures, even light gases like carbon monoxide and methane begin to solidify, settling down onto the surface of the world and becoming integrated into its ice-and-rock matrix. Hardened gases, however, are still just gases, and while they may add to the overall weight of the body, they reduce its overall density. The five large moons of Uranus appear to abide by this rule, appearing to be a bit less densely packed than Saturn's big moons and much less than Jupiter's.

So much volatile gas and ice chemistry, however, ought to make even relatively insubstantial moons active indeed, and at least some of the Uranian satellites fairly surge with energy. Ariel, measuring 718 miles across, is Uranus's brightest bauble, reflecting more than a third of the sunlight that strikes it. The moon's rhinestone reflectivity is a result of its bright, frozen surface, a surface that appears—by cosmic standards at least—to be relatively young. Though Ariel does have craters, many appear to be partially flooded with a now-hardened material, suggesting that something percolated up from underground and filled them after they formed. More curiously, Ariel, like Mars, appears to be scarred by extensive canyons and crevasses, some hundreds of miles long and more than six miles deep. Unlike Martian canyons, which are characterized by jagged walls and rubbly floors, Ariel's have been troweled—almost polished—until they are shiny and smooth. It's likely that some kind of flowing river of liquid was responsible for this natural buffing, but it's unlikely that the liquid was water, since water behaves like steel at Ariel's −335-degree temperature. Instead, whatever rivers existed on the moon were likely made of liquefied methane, ammonia, or carbon monoxide, driven to the surface by the heat of radioactive elements in the moon's core or by residual tidal pumping from the other Uranian moons.

Umbriel, a similarly sized sister moon located one orbit out from Ariel, is, superficially at least, everything its satellite sibling isn't. The darkest of the large Uranian moons, reflecting only 18 percent of the sunlight that strikes it, Umbriel also has one of the oldest surfaces in the Uranian system, pocked with ancient craters that have changed little since the moon's formation. This lack of surface activity is probably due to the moon's comparatively isolated orbit, which keeps it from getting tidally squeezed, but whatever the reason, Umbriel is clearly a static world, one that has experienced none of the upwellings that have so refreshed Ariel.

Oberon, the highest orbiting of the Uranian moons and, at 943 miles in diameter, the second largest, is similar to ancient Umbriel, with a dark, heavily pounded crust that has seen little resurfacing in the last 4.5 billion years. So uneroded is the face of Oberon that *Voyager 2* spotted a four-mile-high mountain on one of its horizons, the equivalent of a thirty-four-mile-high mountain on Earth. Titania, which measures 978 miles across and orbits one slot lower than Oberon, more closely resembles bright, white Ariel, with a newer, polished surface that has clearly benefited from some kind of periodic refinishing.

Similar as these four moons were to one another in size and composition, they were nothing like the smallest and the last of the main Uranian satellites: tiny, fractured Miranda. Measuring just 292 miles across, Miranda is the lowest flying of the five larger moons, and thus lies deeper than any of the others within the Uranian gravity well. The tidal squeezing that this has caused the moon to experience has evidently taken its toll. When *Voyager 2* made its initial Miranda flyby, the fractured surface of the world stopped the reporters in the JPL press room cold. With later refinement—allowing the scientists to make out features as small as a few hundred feet across—the images became more dramatic still. The scientists saw sheer scarps climbing three miles into the sky; cliff faces scarred by parallel, cat-like scratches, apparently created when huge fault blocks rubbed against one another; whole chunks of surface that appeared to be broken and tipped; vast fault lines running for miles through the crust; sudden changes in the moon's reflectivity, with dark, dull expanses lying in a quiltwork pattern next to bright, glassy ones.

Clearly, the *Voyager* team's initial analysis was a plausible one. This was a damaged world, a broken world, a world that had been busted to rubble as many as five times in its past only to shrug off the insult and gravitationally reassemble itself, fusing its shards together in an increasingly chaotic order. Had the breakup occurred only once, the *Voyager* scientists might have attributed it to a chance meteor hit. The fact that it appeared to have happened again and again almost certainly pointed the finger at gravitational pumping, a tidal violence caused by both the tug of giant Uranus lying beneath the moon and the intermittent turbulence caused by Ariel, Umbriel, Titania, and Oberon passing repeatedly overhead. These forces had helped break Miranda up in the past, and as long as they continued to exist, they would likely do so again.

If the moon was indeed going to crumble once more, however, it was going to crumble alone. On January 25, 1986, less than a day after *Voyager 2* arrived in the Uranian system and three days before seven luckless astronauts lost their lives on Earth, the now well-traveled interplanetary ship sped past Uranus, past its eleven rings, and past the last of its fifteen moons. *Voyager 2* would now have just three more years and one more planet ahead of it before it left the solar system altogether. After that, it would have forever.

Late 1986

It was long before *Voyager 2* reached its rendezvous with Neptune that JPL scientists began to realize it might be utterly pointless to make the trip. If the ship didn't collide with the planet, it would all but certainly fail to take any pictures of it; if it somehow took a few pictures, it would certainly fail to beam them home. No matter how rosily the researchers analyzed the journey that lay ahead of their spacecraft, even the most sanguine of them had a hard time believing that the ship's four-planet mission wouldn't end after three.

The biggest thing that stood in the way of *Voyager 2* successfully reconnoitering Neptune was the problem of light. Sitting nearly 2.8 billion miles from the center of the solar system, Neptune receives only the feeblest firefly illumination from the sun and remains lost in what, by Earthly standards at least, is a permanent night. With a little time, a lot of patience, and a lot of acclimation, it might be possible for a human eye to squint through the gloaming and make out a few details of the place. The less sharp eye of a camera would have a far harder time of things, particularly if that camera were mounted on a speeding spacecraft moving at 35,000 miles per hour over the blacked-out Neptunian landscape.

Making matters worse for scientists attempting such an improbable reconnaissance was the route they'd have to fly to get there. In addition to studying Neptune itself, the *Voyager* team was hoping to collect some pictures of the planet's vast, chemically rich moon, Triton. Discovered in 1846, only a few weeks after Neptune itself, the 1,674-mile-wide Triton had been extensively studied since with ground-based telescopes equipped with chemical spectrometers. Though the data such long-distance scanning could gather was imperfect, most astronomers had come to conclude that Triton is fairly drenched in organic chemicals, particularly ammonia, carbon, hydrogen, oxygen, and nitrogen. Circling Neptune in the same kind of highly inclined, backward orbit Phoebe inscribes around Saturn, Triton was almost certainly subject to all kinds of gravitational tugging from its parent planet, which could easily cause the moon's innards to heat up; this, in turn, could cook its organic chemicals together in any manner of interesting ways.

The problem was, trying to fly close enough to Triton to get a good look at it would take some doing. On its current trajectory, *Voyager 2* was following a path that took it just over the top of the solar system, buzzing the north pole of the planets at little better than crop duster altitude. At the moment, however, Triton's sharply tilted, 157-degree orbit placed it far below the Neptunian equator, and thus completely out of the ship's way. Reaching the moon was not impossible, but to get there *Voyager 2* would have to make the same kind of close pass over Neptune's north pole that *Voyager 1* had made over Titan's south pole when it passed through the Saturnian system. Just as the Titanian gravity had flipped the first ship upward, so too could the Neptunian gravity flip the second ship down toward Triton.

As nearly as the trajectory planners could figure it, managing this navigational stunt at Neptune would be even harder than it had been at Titan. In order to bend its trajectory enough to reach Triton, *Voyager 2* would have to pass barely 3,000 miles above the Neptunian cloudtops. When you're making a high-speed approach to a world whose equatorial diameter exceeds 30,000 miles, a margin of error of 3,000 miles is not much. When that world is shrouded in a dense atmosphere that swirls hundreds or thousands of miles into space, things get even trickier. The trajectory plotters at JPL might be able to map the kind of path *Voyager 2* would have to fly to reach Triton, but whether the ship could survive the maneuver was something else again.

Even if *Voyager 2* did manage to high-step its way around Neptune, fly past Triton, and somehow take pictures of both shadow worlds, there still was no guarantee those pictures would ever make it home. While the great, twelve-foot, parabolic antenna both *Voyagers* carried made it easy for the ships to beam information home to Earth, at some point even the best system reaches its operational limits. The strongest transmitter the spacecraft carried sent out a signal no more powerful than 23 watts—less electromagnetic oomph than a refrigerator lightbulb. Around Jupiter, Saturn, and Uranus, even so modest a signal was more than adequate. Around Neptune, however, the beacon became fainter still, spreading and fading until it was barely detectable at all. The answer to the problem lay not in making the transmissions from the spacecraft more powerful, but making the receivers on the ground listen harder.

Throughout the nine years the *Voyagers* had been flying, the JPL controllers had received data from the spacecraft through a trio of large Earth-based antennas arrayed at equal distances around the world. For the Neptune encounter, the JPL team hoped to add a lot of other ears. In Socorro, New Mexico, the space community maintained a very large array of deep-space antennas that had been named, prosaically enough, the Very Large Array. The VLA was made up of no fewer than twenty-eight antennas, each measuring no fewer than eighty-one feet in diameter, and all built to listen not just to the breathy piping of a single spacecraft barely 2.7 billion miles away in space, but to the electromagnetic roar of the cosmos as a whole, the accumulated emissions of countless billions of bodies located countless billions of miles away. If you were somehow able to aim this system to *Voyager 2*, you ought to be able to do a lot of listening, indeed.

The problem was, the *Voyager 2* scientists were not the only ones hoping to lay claim to the VLA. At any given time, cosmologists all over the world were regularly petitioning for a little time with the huge system. With the *Voyager* flyby of Neptune less than three years away and the waiting list a long one, the JPL bosses knew they would never get a turn with the antennas by the ship's 1989 rendezvous date unless they came up with some way to jump the line. They figured they knew just how.

For some time the engineers at the Pasadena lab had been studying the specs of the VLA system, and had come up with a way to redesign its programming guts and dramatically boost its power. If the government and university researchers who ran the VLA gave their approval, JPL would be happy to get to work straightaway on such an upgrade. All the lab asked in return was ninety days of exclusive access to the VLA in July, August, and September 1989, when *Voyager 2* would be approaching, reconnoitering, and leaving the Neptunian system. After the ship had passed through and the encounter was over, the upgrade would be the VLA's to keep. No fools, the managers of the antenna array saw the offer for the sweetheart deal it was and quickly accepted the arrangement.

Figuring out a way to take pictures that the beefed-up antennas could receive was a trickier business. Even in the dim conditions of the Neptunian system, it was possible for *Voyager 2*'s cameras to collect enough light to assemble an image, provided their shutters were left open long enough. In the early days of the MJS77 program the JPL designers had anticipated just this problem and had come up with the image motion compensation technique that allowed the spacecraft to pivot as it flew by a target planet, keeping its eye fixed on a single spot on the world so that the image wouldn't blur. Such a method was used successfully during the Saturn and Uranus encounters, but during those flybys the relatively bright light of the comparatively close sun had made the job easy. Out in the blackness where Neptune spins, the shutter would have to be left open for up to a full sixty seconds to gather enough light for a usable picture, a stretch of time that could challenge even the most deft motion-compensation strategy.

The answer was not to try to capture Neptune or Triton in a single photographic frame, but rather, in a lot of close-up frames. The greater the proximity of the ship to the planet and its big moon, the greater the detail it would be able to make out in the weak reflected sunlight. Taking such extreme close-ups made it impossible to frame either world in its entirety, but if the cameras took a patchwork of partial images—shifting their focus every sixty seconds to capture one part of the terrain, then another, then another—they ought to be able to assemble enough photographic panels for a full, panoramic portrait or two.

That was the theory anyway, and the *Voyager 2* imaging team was convinced it was a good one. The only remaining challenge, after the imagery and antenna problems had been worked out, was the ongoing trajectory problem, and the JPL engineers could not do a thing about that. If *Voyager 2* was going to fly by Triton, it was first going to have to pass within 3,000 miles of Neptune, facing the real possibility that it was going to fly *into* Neptune. The planet's atmosphere, the JPL scientists realized, would either reach out and claim their ship or it wouldn't. Either way, it would be another three years before they knew for sure.

August 25, 1989, 1 A.M.
Pasadena Time

From the surface of Triton you'd rarely get the chance to watch Neptune rise or set. Certainly, you'd *see* the planet. Indeed, it would be impossible not to see it: a dimly lit, ocean blue world measuring 31,000 miles across, lying in space just 220,000 miles away, it would cover fifteen times more sky than a harvest moon covers on an autumn night on Earth. Yet, unlike the moon, it would almost never rise or set.

What Neptune would appear to do is glide. Far ahead of you, hovering low above one of the vast, arctic ice plains that make up Triton, the planet would hang stationary in the sky, its blue disk barely grazing the white line of the horizon. Quickly, then—since Triton speeds around its parent planet at nearly six times the velocity Earth's own moon moves—Neptune would appear simply to drift right, or north, riding the ice prairie like a mammoth sailing ship, only vanishing from view when it had traveled into the distant northernmost latitude, where it would sink briefly out of sight. A short time later it would reappear to the left, or the south, once again hugging the horizon, and once again moving steadily to the right, without ever moving even incrementally up.

It's the sharply skewed angle at which Triton orbits Neptune that causes the planet to appear to follow such an unusual path, but from the surface of Triton it's Neptune that appears cockeyed, lying on its side with its south pole pointing left and its north pole pointing right. It's impossible to say what you'd have seen if you looked at that north Neptunian pole from your vantage point on Triton at about an hour after midnight California time on the morning of August 25, 1989. Certainly, the spacecraft that was soaring over the planet at that precise instant was too small to see from so great a distance. Even if it *was* big enough, it was so close to Neptune that even a single sigh of atmosphere that happened to drift up at the wrong moment would have been more than enough to obscure it completely—and quite possibly to bat it straight out of the sky. On the other hand, the ship was a bright thing—a crinkled clump of shiny gold metal with a higher reflectivity than anything else in the local sky. It was at least theoretically possible that a little of the feeble sunlight that makes it out this far might have been able to catch one of the facets of the spacecraft and flash a quick flicker down to you.

Regardless of whether a person standing on Triton would know the spacecraft barnstorming Neptune existed, countless people 2.7 billion miles away on Earth certainly did. As *Voyager 2* buzzed Neptune that Friday morning and six hours later buzzed Triton itself, it beamed a stream of 23-watt signals back toward terrestrial listeners at a speed of 186,000 miles per second. Even at that velocity the transmissions took four hours and eleven minutes to reach the waiting receivers in Socorro, New Mexico. From there they were relayed an additional 650 miles to the Jet Propulsion Laboratory in Pasadena, where they were translated into pictures. When they were, they told a complicated tale.

After *Voyager 2*'s rendezvous with Uranus three years earlier, JPL scientists did not necessarily expect much from the pictures the ship would return of Neptune itself. It was Uranus's fantastic distance from the sun that likely contributed the most to its unremarkable appearance, with the extreme cold of deep space helping to keep the planet still and quiet, and preventing so much as a breeze from disturbing its heavy quilt of atmosphere. Out around the vicinity of Neptune—a billion miles more distant than Uranus—things should be colder and quieter still.

But Neptune wasn't quiet. When *Voyager 2* engaged its thermal sensors and pointed them toward the planet, it discovered that Neptune actually radiates twice as much heat and nonthermal energy as it absorbs from the sun. Since even a perfectly polished mirror world could never reflect more energy than it receives, Neptune must be generating its own, likely from a dense mass of radioactive elements in its core. While the upper layers of the Neptunian atmosphere are a frigid −360 degrees, this sizzling planetary center causes the lower layers to soar above 900 degrees, and this stirs the atmosphere up pretty violently.

Around the Neptunian equator, gales howl at a supersonic 1,200 miles per hour. The blue face of the planet is streaked by white, wind-blown cirrus clouds, some of which move even faster, speeding around Neptune's entire 96,633-mile circumference in just sixteen hours. The southern hemisphere is stained by a great dark spot as wide as Earth itself, where anti-cyclonic winds blow westerly at nearly 700 miles per hour. In this meteorological maelstrom, atmospheric formations come and go so quickly that in just the thirty-six hours *Voyager 2* was watching, the shredded white clouds that surround the blue spot dissipated almost completely and then were quickly replaced by other ones.

The expanse of space surrounding the planet was just as complex. Earth-based astronomers had long observed the same kind of star-flickering phenomenon in the vicinity of Neptune that they had observed near the other outer planets, and concluded that this last of the giant worlds probably had at least a tenuous set of rings, too. In the case of Neptune, however, those rings appeared to be incomplete, since the stars seemed to wink off at some points near the planet but not others. This led the astronomers to conclude that Neptune is probably decorated not by complete bands, but by arcs, partial circles that simply do not contain enough material to stretch all the way around the world.

The theory was a sensible one—and indeed, judging by the visual evidence at least, was the *only* sensible one—but when *Voyager 2* arrived in the Neptunian system it found that it simply wasn't true. Neptune, it turned out, has no fewer than four concentric rings, all running a full 360 degrees around the planet's equator. While the rings are complete, they are surprisingly sloppy, dense and opaque in some points and transparently faint in others. The lack of tidiness is probably caused by a scattering of large boulder-like masses that are stirred in with the otherwise fine mash of the ring material, causing gravitational clumping that blocks out stars in some stretches of the rings, while allowing them to shine easily through elsewhere.

Equally striking were Neptune's moons. Before *Voyager 2* flew, the astronomical community knew of only two Neptunian satellites: the whale-sized Triton and the much smaller Nereid, a 210-mile moon discovered by Dutch-born astronomer Gerard Kuiper in 1949. As with Uranus, however, given the enormity and distance of Neptune, the JPL scientists always suspected there must be at least a few other small moons secretly circling the planet, and when *Voyager 2* arrived, they found out they were right. Flying through the Neptunian system, the spacecraft spotted no fewer than six previously undiscovered satellites, orbiting the planet from a low of just 30,000 miles to a slightly loftier 73,000 miles.

Of the six new moons, five—Naiad, Thalassa, Despina, Galatea, and Larissa—were little more than cosmic flotsam. Measuring between 36 and 119 miles in diameter, they fly low, stay close to Neptune's equatorial line, and speed around the planet in as little as seven hours. Clinging to Neptune's skirts this way, they have largely escaped bombardment by meteorites, since any incoming projectile would be drawn far more easily to the enormous, gravitationally heavy planet rotating slowly beneath the moons than to the darting, gravitationally light moons themselves. Just how the satellites came to fly in such low orbital slots was not clear, but the fact that their paths follow Neptune's center line so precisely suggests that, like the new Uranian moons, they hadn't just flown crazily in and gotten captured by the planet. Rather, they were probably the remains of a larger moon that accreted from the same primal cloud as Neptune and then broke apart as a result of a random collision. Much more than that, *Voyager 2* was not able to determine. As small as the moons are and as remotely as they were studied, the ship's cameras were able to gather up only a few fleeting pixels of visual information before the spacecraft raced on.

The last and highest flying of the new moons—Proteus—yielded at least a little more information, mostly because of its greater size and gravity. Measuring 257 miles across, it is roughly spherical in shape, but only *very* roughly. Compared to the neatly circular stars, planets, and moons that populate the cosmos, Proteus is an asymmetrical mess. A body this size, with a density as great as Proteus appears to have, almost certainly strains the very structure of the moon itself. Like Saturn's Hyperion, Proteus is probably only a few spoonfuls of matter away from being just large enough to collapse into a sphere. Indeed, with a diameter nearly eighty miles greater than Hyperion's, Proteus holds the distinction of being the largest irregular body, or the smallest near-regular one, in the solar system.

But it was Neptune's pair of better-known moons that yielded the most useful data during the *Voyager 2* flyby. Nereid, slightly more reflective than Earth's moon and slightly smaller than nearby Proteus, had already been known to be an erratically orbiting body, but the spacecraft showed how erratic. The moon sails around the planet in a nosebleed-high orbit, with an average altitude of nearly 3.4 million miles. Its average altitude, however, is not a reflection of how swooping that orbit is. At its lowest, Nereid comes within 841,000 miles of Neptune; at its most remote, it strays nearly 5.9 million miles away. Completing one of these huge, egg-shaped laps takes time, and while a moon like tiny Naiad can make nearly four circuits of the planet in the space of a single Earth day, Nereid takes nearly a year to complete just one. A satellite with so irregular an orbit was almost certainly imported from somewhere else in the solar system, and the JPL scientists were pretty sure they knew where.

In 1951, just two years after discovering Nereid, Gerard Kuiper was conducting some calculations about the total mass of the known solar system and came to the conclusion that at least some of it was missing. Forming bodies as big as the sun and the planets is not an efficient process, and for every clump of matter that gets gathered up and patted down into a star or a world, at least a few scraps get left behind. The overlooked material would likely be the most remote—the primal stuff that stayed out of the gravitational reach of the accreting bodies. Marshaling what he knew about the solar system's collective gravity and mass, Kuiper estimated how much lost matter there should be and where it should be located.

Surrounding the known solar system, he concluded, there ought to be a thin belt of matter reaching more than two and a half times farther into space than the 3.7-billion-mile orbit of Pluto. The matter should be made of the most ancient ingredients—rocky, carbon-rich lumps laced with organic gases and other volatiles, and occasionally covered by fluffy water ice and snow. There was no telling how many objects there might be in this river of debris, but Kuiper guessed that there ought to be tens of thousands, many as big as fifty or sixty miles across and a few much bigger.

More complex than this belt of theorized material—which was quickly dubbed the Kuiper Belt—was another hypothetical formation called the Oort Cloud. First proposed by Dutch astronomer Jan Oort in 1950, the Oort Cloud was essentially the same kind of formation as the Kuiper Belt, but fantastically bigger and fantastically farther away. Surrounding the solar system in a three-dimensional sphere instead of a mere two-dimensional disk, the Oort Cloud is thought to stretch eight hundred to one thousand times the distance of Pluto and, like the Kuiper Belt, is probably made of the icy, rocky residue left behind when the solar system formed. In the case of Oort Cloud objects, however, the material didn't simply hang at the fringes of the solar system while the sun, planets, and moons were accreting. Rather, it was actually in the thick of things, swirling among the nascent bodies as they were condensing. At some point, however, this loose detritus flew by one of the large outer planets at precisely the wrong angle and with precisely the wrong momentum and was gravitationally picked up and thrown. Propelled with all the force a gas giant could muster, the stray material found itself flung to the outermost frontiers of the solar system, where it was too close to escape the gravity of the sun completely, but much too far ever to be a true part of its family of planets and moons.

Not all Kuiper Belt and Oort Cloud objects would remain in such cosmic exile, however. Occasionally, one would stray from the distant swarm and tumble in toward the interior of the solar system. If the rogue was able to fall so far that it actually approached the sun, the solar energy would cause its gases to stream and flare, turning the frozen rock into a common comet, one that would fly in and out of the solar system again and again, as long as its new, irregular orbit survived. Other Kuiper Belt and Oort Cloud refugees would not be able to fly so free. Streaking in through the remote reaches of the solar system, they would have to tumble past Jupiter, Saturn, Uranus, and Neptune. If the trajectory an incoming rock was following brought it too close to one of the big worlds, there was always a chance it might get stuck on its gravitational flypaper, abandoning its original path and entering a permanent orbit around the planet.

That, the JPL astronomers figured, was probably what happened to Neptune's little Nereid, whose elliptical orbit looked like a last gravitational artifact of a violent capture. More important, it was also what probably happened to the planet's giant satellite Triton. The second most highly inclined moon in the solar system—behind Saturn's Phoebe—Triton not only circles Neptune at a 157-degree angle, but also clockwise, moving in a direction opposite almost all of the solar system's other worlds. An orbit like this suggests not only that Triton was captured by Neptune, but also that that capture was even more violent than Nereid's. The harder the gravitational brakes were slammed on, the more skewed Triton's orbit was likely to be, and the more skewed the orbit was, the more tidal energy the moon would likely absorb. This would have generated a lot of heat, meaning that Triton ought to be a warm place.

When *Voyager 2* turned its thermal sensors toward Triton, however, it found just the opposite. Whatever tidal energy the moon might once have radiated had apparently been largely spent, meaning that Triton was not only not warm, but paralyzingly cold—about −391 degrees, as far as the spacecraft could tell. This qualified Triton as the coldest object in the known solar system. Most of the frigid moon is covered by a solid permafrost of nitrogen, carbon dioxide, carbon monoxide, and methane—materials that would normally vaporize at the barest breath of solar warmth, but out in the Neptunian neighborhood, harden into icy armor. This frozen skin shines so snowy a shade of white that Triton as a whole reflects roughly 70 percent of the sunlight that strikes it. So far from the center of the solar system, that is not much sunlight at all, but if Triton were somehow made to orbit Earth—and were somehow able to stay frozen while it did—it would shine nearly six times brighter than a full moon.

Cold as Triton's exterior is, however, residual tidal forces do appear to have an influence on its innards. The moon as a whole is largely free of craters, indicating an outward-radiating heat that has repeatedly melted and resurfaced the crust. Fine cracks run like webwork through some parts of the surface, suggesting the same kind of thermal expansion and contraction that fractures Jupiter's Europa. Elsewhere, there are numerous glass-smooth plains, calling to mind the seas on Earth's moon. Unlike those lunar basins, however, which were smoothed over by molten rock seeping up from underground, Triton's seas were likely formed by a cold ammonia and water slurry that oozed slushily up from the warmer interior.

Of course, just when all this crustal sloshing and shifting took place is a mystery, and it's possible that recent epochs have gone by without much fresh resurfacing on Triton at all. But even if most of Triton's tidal warmth has dissipated, the Neptunian satellite is not a dormant place. As *Voyager 2*'s initial images of Triton came back, JPL scientists saw features that they could swear looked like active volcanic geysers. Here and there over the ice white surface of the moon there appeared to be tall, black, feathery plumes, rising so high in the sky they even cast shadows on the ground below. Elsewhere, the plumes and shadows were gone but a distinctive black streak remained on the ice—the ghostly smoke of an earlier geyser that had settled back to the surface, leaving a telltale stain.

As tempting as it was to conclude that the thermal engine behind these blasts of icy exhaust lay in the tidal energy of the planet, the *Voyager* astronomers bet that the likeliest answer was the sun. Even in Triton's deep-space neighborhood, at least a shimmer of solar energy does manage to survive. When this faint illumination strikes Triton, it easily penetrates the nitrogen ice that covers its surface and bathes the black, carbon-rich layer underneath. Like all dark materials, carbon absorbs light and heat—even the impossibly weak light and heat from a sun more than 2.8 billion miles away. As the carbon grows incrementally warmer, the frozen nitrogen lying directly atop it flashes into gas and begins migrating beneath the overlying crust looking for a weak spot. When it finds one, it blows through the icy shell and explodes into the sky, carrying clouds of sooty carbon with it. After the blast of gas fully expends itself, the carbon settles back down to the ground and the crack closes over.

Though Triton's geysering has likely gone on for billions of years, it is unlikely to last billions more, largely because the moon itself is probably doomed. Planets and their satellites generally try to achieve a gravitational balance, and one of the things that throws that balance off is a retrograde orbit. When a moon orbits clockwise like Triton and a planet rotates counterclockwise like Neptune, the constant gravitational drag that results causes the moon to lose energy. As this happens, the satellite drops lower and lower in its orbit until—after millions upon millions of years—it simply falls out of the sky and collides with the planet below. Triton is now undergoing just such orbital decay and is all but certain to come to just such a calamitous end. The only thing that could save it will be if it plunges in at such an angle and such a speed that Neptune's gravity tears it apart before it completes its fall, turning it into a cloud of rock and ice that disperses around the world, forming a grand and gaudy set of rings. No matter how Triton meets its end, it is sure to die. And it will be Neptune itself, whose total number of known moons so recently grew to eight, that will be responsible for taking its first daughter's life.

•   •   •

On August 25, 1989, before the California morning had passed to afternoon, the Jet Propulsion Laboratory in Pasadena received a signal confirming that *Voyager 2* had passed the last of Neptune's moons, dropped out of the bottom of the solar system, and begun its infinite fall into interstellar space. In twenty-three years it would reach the heliopause, that point past which the effects of the charged particles in the solar wind are no longer felt. In 24,000 years it would reach the Oort Cloud. In 294,000 years it would make its first comparatively close approach to a star other than the sun, flying by the bright solar furnace of Sirius. Within a million years it would pass a dozen more stars.

Where *Voyager 2* will go after that is more difficult to chart, but long before it left the ground the JPL designers decided to plan for its journey. Attached to the side of both *Voyager*s is a twelve-inch gold-plated disk, etched with analog grooves like an ordinary LP record. Protecting the disk is an aluminum cover engraved with diagrams explaining in mathematical symbols the origin of each spacecraft, its basic trajectory, and a rough idea of when it was launched. Additionally, there are a few schematic scribbles explaining how to operate a tiny stylus stored beneath the cover.

Should anyone find the spacecraft, apply the stylus to the grooved disk, and somehow know to connect the whole apparatus to a turntable, a speaker, and a video system, the golden record would reveal a lot. Encoded in its grooves is a portfolio of color images from Earth, including seashores, snowflakes, dolphins, eagles, a mother nursing a baby, a house under construction, a fishing boat spreading its nets, and a gymnast spinning through her performance. Also included are sounds of the wind, the rain, and the surf; a cricket chirping, a frog croaking, and a horse clopping; a tractor, a riveter, a train, a bus, a Saturn 5 rocket leaving the ground, and spoken greetings in dozens of languages. Finally, the disk includes music—Mexican music, Javanese music, Peruvian music, Chinese music, Japanese music, Senegalese music; there is a little bit of Bach, a little bit of Mozart, a little bit of Beethoven, a little bit of Chuck Berry, and a complete recording of "Melancholy Blues," by Louis Armstrong and his Hot Seven Band. In all, there are 116 pictures, thirty-nine sounds, fifty-five samples of spoken languages, and twenty-seven selections of music.

It is virtually certain, of course, that the music will never be heard, the pictures will never be seen, the language will never be listened to or fathomed. Instead, the ships are probably destined to wander forever and alone, journeying indefinitely through the vast wilderness of the galaxy, without ever being noticed, much less intercepted. In the laughably improbable event that they are one day found, however—in the event that they are somehow captured and the code their records carry is cracked—the *Voyager* spacecraft could serve as one of humanity's most profound statements. There is some satisfaction, after all, in boasting to the cosmos that a civilization that could build a rocket and launch a space probe could also produce a Louis Armstrong.

The Deep Solar System

April 1998

If you wanted to hijack the *Voyager* spacecraft, you'd have a pretty easy time of it. Certainly, making contact with the two ships would be no simple matter; deciding what to say to them after you finally did get in touch would be a tricky business, too. But before you could communicate with the distant probes, you'd have to figure out where they were, and that at least would be a snap.

Where you'd find the two *Voyager* ships—or where you'd find the lonely listening post that would allow you to talk to them at least—would be in an unremarkable redbrick building on an unremarkable stretch of Madre Street, in Pasadena, California. Up the road a bit, just one or two clicks to the north, you'd find a little strip of shops—a grocery store, a coffeehouse, a magazine stand, a pet shop. Off in the other direction, about the same number of clicks to the south, you'd find the turnoff to Interstate 210, which would take you to the Jet Propulsion Laboratory. You'd know you were approaching JPL even if there weren't any signs to tell you so, mostly because of all the obstacles that prevent you from actually getting there. The sprawling facility is protected on one side by a forbidding stretch of patchy green mountains and on the other side by fences and guard booths and a series of round-the-clock checkpoints—just the kind of impenetrable perimeter you'd expect around so glamorous a lab doing such glamorous work.

There was a time when the *Voyager* listening post enjoyed the same kind of protection. But then, there was a time when the people behind the JPL barricades were fiercely concerned with what happened to the pair of ships. Every morning hundreds of scientists and support workers were waved past the guard stations and cleared into the buildings, and there spent the entirety of their days seeing to it that the ships kept flying, their instruments kept working, and the planets and moons they were approaching were the ones they were supposed to be approaching. But that was a lot of years and a lot of planets and a lot of cosmic miles ago. Things are different now.

If you wanted to reach the *Voyager* spacecraft in the last few years, what you'd have to do first is reach Richard Rudd. Rudd is the mission director of the *Voyager* program—maybe the eighth one, maybe the ninth one, maybe the eleventh one; people stopped paying close attention to the exact number after the whole project was bundled up and moved to its redbrick building on Madre Street. Not that the *Voyager*s don't still turn in a good day's work. *Voyager 1*, which hasn't been within hailing distance of a planet or moon since it soared up and out of the solar system in the autumn of 1980, is now 6.51 billion miles from Earth, or about twice as far from the sun as Pluto is and seventy times more distant than tiny Earth. *Voyager 2*, which fell out of the solar system after plunging by Neptune nearly a decade ago, is about 5.02 billion miles away, or fifty-four times the Earth-sun distance.

Out in that cosmic wilderness, the spacecraft's magnificent cameras don't get much of a workout, and it has been more than eight years since they sent home so much as a single grainy image. The final picture a *Voyager* did return, however, was a dandy. In 1991, eleven years after *Voyager 1* completed its primary mission, JPL scientists made an unexpected discovery. Calculating the positions of all nine planets and comparing them with the position of the spacecraft perched high above the solar plane, they realized that if the ship's cameras were aimed precisely enough and its shutters were held open long enough, it just might be possible to take the first-ever family portrait of the entire known solar system.

The imaging team told the engineers just how the picture should be taken, the engineers sent the instructions up to the ship, and over the next several days the ship did its best to execute the command, turning and firing, turning and firing, trying to herd all nine planets into a single photographic frame. As it turned out, it wasn't possible. Tiny Mercury, the innermost of the planets, was all but entirely washed out by the glare of the nearby sun. Mars, a bit larger and a bit farther from the solar system's center, was lost in shade. Pluto, spinning in its cosmic wilderness, was shrouded in shadow as well. Nonetheless, Venus, Earth, Jupiter, Saturn, Uranus, and Neptune were all visible, and all did sit still for the camera, appearing in the frame as six tiny points of light freckling a canvas that was otherwise a pure, glossy black.

The extraordinary photo was quickly processed into first dozens, then hundreds of copies and distributed around the JPL campus. Scientists tacked the picture up on their bulletin boards, taped it onto their office doors, took it home to show to their families. And then, almost by consensus, they forgot about it. There were other missions to be flown, after all, and newer ships to be built—ships that would have more serious work to do than fooling about with something as frivolous as a planetary family picture—and it was those projects that deserved the lab's attention now. By the time the second smallest of the points in that final *Voyager* photograph—the faintly blue one located not far from the sun—had completed one more rotation, the JPL engineers instructed the two ships to close their photographic eyes for good.

Nonetheless, even now, at the century's end, *Voyagers* 1 and 2 do still communicate with home—and quite a bit, in fact. Anywhere from ten to sixteen hours per day, the now-blind ships continue to speed through space, sniffing the void around them and sending home data on the fields and particles, solar wind, and magnetic fields they find there. In all, seven sensors continue to operate on both spacecraft, though the plasma detector on *Voyager 1* has worked only spottily for years, having been scalded by Jupiter's radiation fields when it flew by the planet in 1979. The transmissions that do come back from the *Voyagers* generally come back to Richard Rudd, who collects and collates them, and studies what they have to say to him. He knows, however, that what they do tell him is not likely to make many headlines. So routine are the ships' incoming signals and so unhurried are the doings in the lab that now and then the cubicle workstations outside Rudd's office door—the stations where a small team of flight controllers ought, in theory, to be sitting—are left completely unattended. To prevent the spacecraft from falling into careless or mischievous hands, some of the mission scientists have been careful to glue bright orange paper warnings to the keyboards and monitors of their unsupervised computers.

"CAUTION," the little stickers say. "THIS IS A LIVE VOYAGER CONSOLE. DO NOT TOUCH." It is here you could come to talk to the *Voyager*s if you ever had a mind to. And it is here, if you had the skills and the inclination and a slightly criminal bent, that you could try to shanghai the ships. The catch, of course—and this the JPL engineers know—is where could you try to send a spacecraft that's already nowhere at all?

The *Voyager*s thus continue to fly largely unremarked upon, telling their caretakers on the ground little that's scientifically newsworthy. Even if the control center on Madre Street is a sleepy one, however, that doesn't mean JPL engineers as a whole have nothing to keep them busy. Elsewhere in the cosmos, as it happens, other spacecraft are saying plenty.

For all the worlds the *Voyager*s toured, it was Jupiter that most demanded a return visit. It wasn't just the planet's vast size that captivated the scientists; it wasn't just its diaphanous rings or its swarm of moons. It was, ultimately, all of those things. A planetary system so dense and complex could not possibly be fathomed by a *Voyager* spacecraft blowing by at 35,000 miles per hour and then vanishing forever into deeper space. What was needed was a bigger and more versatile ship, one that wouldn't just reconnoiter the world but orbit it, survey it, dodge and dart among its moons, and perhaps even taste the atmosphere and chemistry of the planet itself. No sooner had the *Voyager*s completed their own Jovian flybys in 1979 than the JPL engineers began working on just such a souped-up space machine. They would call their spacecraft, fittingly enough, *Galileo,* and designing it would present them with challenges like none they had ever encountered before.

The problem with trying to put a spacecraft into orbit around a world is that that spacecraft has to be a big one. In order to leave the straight, ballistic path the *Voyagers* flew and instead inscribe an orbital circle, an interplanetary ship must first slow down, bleeding away enough speed and energy to allow the planet's gravity to catch and hold it. That kind of deceleration requires a powerful engine—one made up of a lot of machinery and carrying a lot of fuel; that, in turn, means a lot of weight. The *Voyager* spacecraft were relatively light, relatively lean machines, tipping the scales at just 1,793 pounds each. The *Pioneers* were a flyweight 568. *Galileo*, the JPL engineers figured, would need to weigh close to three tons.

More difficult than the job of building so big a ship would be the job of flying it. If mission planners were serious about trying to sample the Jovian atmosphere, they were going to have to come up with a way to bring their spacecraft into direct contact with the planet's gases—and that could spell trouble. Flight planners designing trajectories for Earth-orbiting spacecraft know that even the most tenuous wisps of high-altitude air can be fatal to a ship, dragging on it until its attitude destabilizes, its orbit decays, and it goes plunging toward the ground like a brightly flaming meteor. A far faster interplanetary spacecraft encountering the far denser atmosphere of Jupiter would be annihilated on contact. In order to prevent this from happening, the JPL engineers would have to build a *Galileo* spacecraft that was essentially two spacecraft.

Stowed aboard the ship before it was launched would be a four-foot-wide, three-foot-long bullet-like probe, packed with half a dozen sensing instruments, including a wind-speed monitor, a chemical spectrometer, a lightning detector, a particle sensor, and a battery to power them all. During the years the spacecraft was making its transit to the Jovian system, the probe would do essentially nothing at all. Five months before arrival at Jupiter, however, *Galileo* would fire the little instrument package into space and then, with a puff from its thrusters, nudge itself off on a slightly different trajectory from that of its now-jettisoned cargo. Over the 150 days that followed, the main body of the *Galileo* ship would continue on a broad, arcing path that would take it obliquely toward the planet, where it would fire its engine and ease gently into Jovian orbit. The probe, by contrast, would aim itself for a bull's-eye strike just above the planet's equator.

When the ballistic projectile did hit Jupiter, it would hit hard, colliding with the atmosphere at nearly 100,000 miles per hour. Punching through the air, blunt end first, it would plunge for a full two minutes, until atmospheric friction had slowed it down to less than a hundredth of its original speed—slow enough to allow it to deploy a parachute. If the big cloth canopy successfully opened—by no means a sure thing—the probe would immediately switch on its sensors and begin sending what it learned about the atmosphere around it back to Earth. For the next hour and a quarter the probe would operate, dropping about 125 miles through the orange-red clouds, while the air pressure around it rose from almost nothing at all to twenty-five times the pressure found at sea level on Earth. Finally, the crush and heat of the atmosphere would grow so great that the probe would simply crumple, melt, and ultimately vaporize. Just over seventy-five minutes after arriving at Jupiter, the exquisitely engineered instrument package would be reduced to mere superheated vapor, adding its own tiny breath of largely metallic gases to the hydrogen and helium world it now inhabited. High above, the surviving mother ship would spend the next two years orbiting the planet alone.

That was the plan at least, but executing it—and paying for it—would be no small matter. It took more than a decade and more than a billion dollars before *Galileo* was actually engineered, built, and readied for flight, and it was not until October 1989 that it actually left its Florida launch pad—setting out on a slow, looping trajectory that would not get it to Jupiter until December 1995. But before *Galileo* could travel even a quarter of the way to the planet, it would become clear to JPL engineers that they were flying a badly snakebit ship.

The first of the spacecraft's problems concerned its high-gain antenna. Like the *Voyager* ships, *Galileo* was equipped with a sensitive antenna dish that would allow it to communicate with Earth from distances of hundreds of millions of miles. The *Voyager* antennas had been impressive assemblies, measuring about twelve feet across. For *Galileo,* the JPL engineers hoped to improve communications even further, equipping the new ship with a larger high-gain system measuring a formidable 15.75 feet. But while the bigger system was, indeed, a more powerful one, transporting it into space presented problems.

The *Voyager*s were fired on their interplanetary path by a Titan-Centaur booster, a stout, two-stage rocket with a roomy cargo bay more than big enough to handle the oversized high-gain antenna. *Galileo,* by contrast, was designed to be flown into Earth orbit inside the cargo bay of a space shuttle and then released into space to make its way to Jupiter on its own. A shuttle's cargo bay, however, is a cramped place, hardly big enough to accommodate the old *Voyager* antennas, much less *Galileo*'s larger one. The answer was to reengineer the *Galileo* antenna entirely. While the *Voyager* dishes were molded from epoxy into a single rigid piece, the *Galileo* high-gain antenna would be completely collapsible. Built from eighteen metal ribs and covered in a flexible gold mesh, it would fold up like an ordinary umbrella before being packed away in the shuttle. Later, after the ship had been set free from the orbiter and fired on its trans-Jupiter path, a signal would go up from the ground releasing the ribs, unfurling the antenna, and at last giving the spacecraft its electronic voice.

Precisely when that antenna deployment would take place presented problems of its own. In order to build up the speed it needed to fly all the way out to Jupiter, *Galileo* would not follow a straight, as-the-crow-flies path to the planet. Rather, it would be sent on a sort of corkscrew trajectory through the inner solar system, flying around and around the sun in an orbit that would take it once past Venus and twice past Earth. Each time the ship approached one of these large worlds, it would receive a gravitational kick that would add hundreds or thousands of miles per hour to its speed. Those velocity boosts would cause *Galileo*'s orbit to grow wider and wider, until finally, after six years, it would have spiraled all the way out to Jupiter.

Appealing as this fuel-efficient trajectory was, it presented some problems. If mission controllers unfurled the giant high-gain antenna when the ship was still moving through the inner solar system, there was a risk it could be damaged beyond repair as the reflective gold mesh absorbed the full, wilting heat of the nearby sun. To prevent such a mission-ending meltdown, JPL planners decided that *Galileo* would keep the skirts of its high-gain demurely furled for the first eighteen months it was in space, doing all its communicating with Earth through a far smaller low-gain antenna—one able to communicate at a rate of just eight data bits per second. Only when the ship was past the orbit of shadowy Mars would the far larger high-gain be brought on-line. Conversing with *Galileo* at the eight-bit rate the low-gain antenna would allow would be a painstaking business—a little like trying to have a conversation with a person who had all at once lost the ability to speak complete words and was instead reduced to spelling them out. But since such a plodding dialogue would last little more than a year and a half out of an eight-year mission, and since the alternative—exposing the high-gain antenna to the full ferocity of the close-up sun—was out of the question, mission designers figured they'd simply make do.

Even so redundant a pair of antennas, however, left at least a few engineers feeling skittish. While the low-gain antenna would be a perfectly adequate backup system during the lazy months *Galileo* was circling through the inner solar system, when the ship was actually orbiting Jupiter, things would be different. Pictures and other data coming back from the spacecraft would stream directly from its cameras and sensors, through its high-gain antenna, and back to Earth at a rate of thousands of bits per second. If the high gain should fail even briefly during a critical moment, the low-gain could never hope to keep up with this pace, and the data would simply be lost. In order to prevent this from happening, the designers decided to add one more piece of hardware to the ship: a tape recorder. In the event of a temporary high-gain breakdown, data would be stored on the recorder and later played slowly back to Earth through the far slower low-gain system. Conducting the entire mission this way, of course, would be impossible, since even with the tape recorder help-ing out, the low-gain antenna could never transmit all the pictures *Galileo* was built to collect. As a short-term fix, however, one that would be used only until the high-gain antenna was working again, this backup system was a good one.

With such triply redundant hardware built into their spacecraft, mission planners could feel confident that they had designed a com-munications system that was almost completely resistant to failure, and for the first eighteen months after *Galileo*'s 1989 launch, there was no reason to assume anything *would* fail. Finally, on April 11, 1991, when the ship's trajectory had spiraled out as far as the edge of the asteroid belt between Mars and Jupiter, JPL planners decided it was at last probably safe to unlock the high-gain antenna and spread its ribs. It was only then that they'd learn if triply redundant was re-dundant enough.

Though the deployment of the high-gain system was not a com-plicated exercise, it was a critical one, and for that reason the chief-tains of the *Galileo* project made sure they were there to watch it happen. On hand at the flight director's console that afternoon were mission director Neal Ausman, deputy mission director Matt Lan-danow, and project manager Bill O'Neil. O'Neil and Ausman were far and away the higher ranking of the three men, but Landanow, they all knew, was far and away the most knowledgeable. As chief engineer during the *Galileo* design phase, he had familiarized him-self with every strut, nut, and rivet of the ship, and could practically describe their placement and purpose from memory alone. If any-thing went wrong this afternoon, Landanow would likely be the first person to recognize it—and the first person to come up with a way to fix it.

For the first forty minutes or so after the deployment command went up, O'Neil, Ausman, and Landanow had little to do. Like so many other JPL controllers before them, they knew they would have to tolerate the nonnegotiable limits of light speed, waiting twenty minutes as their signal traveled from Pasadena to the spacecraft and then another twenty minutes as it traveled back again. For that entire time their screens told them nothing, flickering merely with the self-evident information that their command had indeed been sent. Finally, after just over the anticipated forty minutes had elapsed, a column of numbers began to blink on the glass. Landanow gave the figures a quick scan and immediately noticed something amiss. He read them again—a bit more closely—and this time started to feel downright queasy. The antenna, from all indications, was pulling what the engineers called stall current. The motor was drawing power, the deployment gears were engaged, but the ribs of the umbrella appeared to be going nowhere at all.

"We're stuck," Landanow said flatly.

"How can you tell?" O'Neil asked.

"The current is saturated, something is jammed," Landanow said. "In any event, the antenna's not budging."

Ausman gave the numbers on the screen a read of his own, confirmed what Landanow was saying, and immediately called out to his flight controllers, instructing them to send a second deployment command up to the ship. The engineers complied, and forty minutes later another stall signal came down. A third command yielded a third signal, and a fourth a fourth. With each new report Landanow winced. If he knew this ship—and he surely did—he could all but guarantee that whatever was hanging up the antenna was not much: a single too-tight fitting, perhaps, a single protruding bolt, one that was situated in just such a way that it managed to jam all eighteen ribs. If it were somehow possible to transport the *Galileo* spacecraft to a hangar in Pasadena, Landanow knew he could probably roll over a stepladder, climb up to the antenna, and spring it free with his hands alone. But *Galileo* was not in a hangar in Pasadena; it was tens of millions of miles away, at the edge of the asteroid belt between Mars and Jupiter, and more elaborate measures would be necessary.

The first fix Landanow recommended was what the engineers referred to as hammering the motor. Sending signal after signal up to the ship, mission controllers would switch the antenna engine on and off at a speed calculated to cause the greatest vibrational resonance possible through the ribs. Anything that was stuck ought, in theory, to become easily unstuck. What the antenna ought to do and what it did do, however, turned out to be two different things, and no matter how many times the engineers hammered the engine throughout that day and the next one and the one that followed, nothing budged.

Making matters more frustrating still, as additional telemetry streamed down from the spacecraft and additional engineers had a chance to analyze it, it became clear that it wasn't the entire antenna that was jammed at all, but just a portion of it. On one side of the ship was a sun sensor that was bathed in solar light when the antenna was closed but ought to fall into shadow when it was opened; according to the data, the sensor was indeed now mostly shaded—but not entirely so. Working with a fully functional stand-in *Galileo* kept in a JPL clean room, the engineers positioned and repositioned the earthbound ship's antenna ribs, trying to reproduce the light and shade signals that were coming in from space. In the configuration that worked best, only three of the spacecraft's eighteen ribs were stuck in the closed position.

But a mostly open antenna was as bad as no antenna at all, and if the mission was going to proceed, the engineers would have to come up with a way to spring the rest of it free. If there was anything they had on their side, it was time. It would be another four years before *Galileo* finished inscribing its lazy arc to Jupiter, and as those years went by, flight controllers continued to try to fix what ailed their injured high-gain: They hammered and rehammered the balky deployment motor, studied and restudied the telemetry streaming back from the ship. Activating *Galileo*'s main engine, they shook the spacecraft this way and that, hoping to jolt the stuck ribs loose. Rotating the ship in toward the sun and then out toward space, they alternately heated and cooled the antenna, hoping the repeated expansion and contraction would pop something—anything—free. But month after month—and, ultimately, year after year—the condition of the high-gain antenna remained unchanged.

Finally, in the middle of 1995, as the spacecraft was approaching its early December arrival at Jupiter, the JPL engineers threw up their hands. The *Galileo* Jupiter ship—the billion-dollar machine the flight directors had spent more than a decade designing, building, and planning to fly—had been struck permanently mute. The ship's cameras and sensors were as alert as ever, its electronic brain was as sharp as ever, but the rare and perishable science that the instruments were built to collect would remain forever imprisoned within the ship.

Or at least it looked as if it would. If the booming voice of the high-gain antenna had been silenced, however, there was always the piping voice of the low-gain, and that provided some hope. As the plans for *Galileo* now stood, the orbiter portion of the spacecraft would be sending home a whopping load of photos—thousands, even tens of thousands, as it cruised around Jupiter and coasted among its moons. Indeed, so enormous would *Galileo*'s picture portfolio be that some of the images would even act as de facto movies—sequential, kinescopic snapshots that, when stacked up and riffled like a flip book, would reveal the motion of Jovian clouds, the pluming of volcanoes on the surface of Io, or maybe even the gush of a warm-water geyser as it hissed up through the ice of Europa.

Suppose, however, the flight controllers were willing to settle for fewer pictures? Suppose they instructed the cameras not to shoot indiscriminately at anything they saw, but to hold their fire until the most photographically opportune moments—when *Galileo* was cruising directly over a moon, say, or directly past the planet. The comparatively few pictures the spacecraft took could then be stored safely on the tape recorder, and later, in the weeks or months it would take the ship to make its next close pass by the next big moon or atmospheric formation, they could be played back to Earth through the slowpoke low-gain antenna. The process would be an excruciatingly tedious one, and it entirely ruled out luxuries like volcanic and atmospheric movies. What's more, it put a colossal burden on a fragile, fallible tape recorder that was intended only for the most occasional use and only in the most critical emergencies. But if the utter loss of the high-gain antenna did not constitute such an emergency, nothing did, and if the tape recorder was going to buckle under the weight of its new workload, that was simply a risk the engineers would have to take.

Their hands effectively tied, the *Galileo* flight directors approved the scaled-back mission, and as the spring and summer of 1995 gave way to autumn, and *Galileo* moved within eight weeks of its scheduled arrival at the planet, they prepared to implement the new plan. What nobody at JPL counted on was that the tape recorder was about to break down, too.

On October 11, 1995—four and a half years to the day after the *Galileo* antenna failed—Landanow, O'Neil, Ausman, and other mission supervisors were gathered in a meeting room in JPL's Building 264 reviewing the ship's upcoming insertion into Jovian orbit, when a member of the engineering team called on the phone. Strictly speaking, members of the engineering team should not be taking it upon themselves to phone any room in which so many project elders were gathered, and if one did, it could only mean that he was calling with very, very good news or very, very bad news. From the tone of the engineer's voice alone, Landanow could tell that this news was very, very bad, indeed.

"We've got a problem with the tape recorder," the engineer said.

"What kind of problem?" Landanow asked.

"It's spinning. We sent it a command instructing the reels to turn and they did turn, but the tape itself didn't go anywhere. Now the reels won't stop."

"Has the tape broken?"

"We don't know."

"Has it spewed off the reels?"

"We don't know."

Landanow rubbed his face. "Well," he said, "I guess we'd better find out."

Landanow adjourned the routine meeting that was under way and within the hour convened a far more urgent one, this time with the members of the flight engineering team. The group's general objective, of course, would be to diagnose the tape recorder's problem and see what they could do to fix it, but their more immediate concern was simply to bring the machine's madly whirling reels to a stop. As it turned out, they were able to accomplish this initial goal relatively easily. Sending up a series of override codes, they succeeded in disengaging and powering down the tape recorder entirely, rendering it temporarily useless, but at least keeping it from doing itself any further harm. Next, they set about analyzing the telemetry readouts to see what had caused the breakdown in the first place.

From what the computer records of the past few hours indicated, the tape recorder's difficulties began when the ground controllers sent it a signal commanding the system to move in reverse, rewinding a short stretch of tape onto the feeder reel. The on-board machinery started to execute the simple command, but according to one curious data blip, it had to overcome a slight resistance first, almost as if the tape had become stuck on something and needed to be pulled free. When it finally *was* freed, the entire assembly simply went slack.

Without opening up the tape recorder and looking inside, there was no way of knowing what had caused the assembly to get hung up this way, but there were only three things that were likely to be responsible. The recording tape itself was manufactured with a chemical binder that helped give it its toughness and resistance to snapping. In the punishing temperatures of deep space, it was always possible that the chemicals on one patch of tape had undergone adhesive changes, causing them to become soft and gummy and to glue the tape in place. In the alternative, it was possible that the aluminum dioxide that made up the machine's recording head was accumulating an electrical charge, and as the tape passed by, it was getting stuck in place, much the way light, loose objects get stuck to a sweater crackling with static electricity. Finally—and more worrisomely—it was also possible that a small barb of metal had loosened itself somewhere in the guts of the machine and was jabbing at the tape as it spooled past, steadily weakening it and threatening to tear it in two. While this last alternative was clearly the most ominous, it was also the least likely, since if the tape was indeed being poked and damaged, it probably would have torn by now.

Whatever was causing the problem, the *Galileo* team eventually learned how to get around it. Putting the tape recorder through some careful, experimental spins, they discovered that—as the initial telemetry signals had suggested—the system almost always worked fine when the reels were moving forward; it was only backward that presented difficulties. In order to overcome this and get the reels going in reverse when they needed them to, the engineers discovered that all they had to do was advance them a little bit first. This somehow seemed to cock the spring-driven system and set it spinning in the other direction, much the way a smart tug downward on a common window blind gets the whole assembly flap-flap-flapping upward. Such a crude fix was not the solution the fussy engineers would have picked if they had had a choice, but the tape recorder had not offered them a choice, and if they wanted the balky system to function again, they would have to change the mission protocols to include this extra step. In the fall of 1995, the rules were indeed rewritten this way and the tape recorder went back into operation.

Several weeks later, on December 7, 1995, at 10:04 P.M. Pasadena time, the *Galileo* atmospheric probe, which had earlier been released from the main body of the ship, plunged into the rusty red Jovian clouds and began sending its stream of data back to the mother ship. Nearly two and a half hours later, at 12:27 on the morning of December 8, the spacecraft itself—its antenna arthritically frozen, its tape recorder cautiously turning—swung behind the great bulk of Jupiter and settled into orbit around it, adding a tiny metal moon to the sixteen natural ones that already circled the world.

For the probe, of course, the day it arrived at the planet was a short one. After entering the atmosphere and deploying its parachute, the 746-pound bullet died even faster than the mission designers had anticipated it would, surviving only fifty-seven minutes and thirty-six seconds before it was reduced to vapor by the crush of the Jovian air. In that brief time, however, it was able to switch on its instruments and drink in a gush of data from the environment around it, beaming it all up to the orbiter, which dutifully preserved it on its tape recorder until all transmissions from the probe stopped. Over the course of the next four months, the orbiter slowly dribbled the data back to Earth through its low-gain antenna. Though the probe's findings were a long time in coming, they turned out to be worth the wait.

Planetary scientists had always assumed that the uppermost reaches of Jupiter's atmosphere, where the air was vanishingly thin and the pressures were almost unmeasurably low, would also be paralyzingly cold—perhaps as low as −300 degrees. As it turned out, however, the temperatures in the region were about 100 degrees warmer than that. This suggested that the planet has some kind of internal heat source apart from its own intense gravitational energy, and that source, the scientists concluded, is probably a vast lode of radioactive material buried deep in the center of the world. Scientists studying data returned by *Voyager 2* had theorized just such a planetary oven at the heart of Neptune, and concluded that this could help explain the planet's high winds. Jupiter's winds were not quite so fast as Neptune's, but the planet did turn out to be a gusty place, with the *Galileo* probe detecting gales of 330 miles per hour in the upper atmosphere and 450 miles per hour down lower. If anything could drive those kinds of air bursts, it was likely buried radioactive heat.

The chemistry of the atmosphere was a bit less surprising. As scientists had anticipated, the Jovian air turned out to be about 13.6 percent helium and the remainder hydrogen—or about the same concentrations as the sun. This helped confirm the idea that Jupiter indeed swirled into existence from the same raw, gaseous material that formed the stars, and suggested that the planet might indeed have been a star itself if only it had been a little larger and a little denser. Simple organic elements like carbon, sulfur, and nitrogen are stirred into the atmospheric mix as well, but more complex hydrocarbons are largely lacking. This absence of preorganics is probably due to a paucity of water and oxygen in the Jovian stew, as well as the relative rarity of lightning, which can play a role in jolting hydrocarbons into existence. As the *Galileo* probe descended through the Jovian air, it did detect the occasional lightning bolt, but the closest one was a good 8,000 miles away.

Illuminating as the findings from the probe were, it was the data from the *Galileo* orbiter that provided the true Jupiter news. One of the least expected findings concerned not the planet itself but its rings. Ever since the *Voyagers'* Jovian flybys in 1979, planetary scientists had been assuming that the fine bands surrounding Jupiter, Uranus, and Neptune were the remains of small, annihilated moons, ones that were destroyed by collisions and then gravitationally prevented from pulling themselves back together. At least in the case of Jupiter, however, *Galileo* images revealed otherwise. Photographing the small inner satellites Metis, Adrastea, Amalthea, and Thebe, the spacecraft discovered that the Jovian moons appear not merely to orbit in the vicinity of the rings as scientists had previously thought, but rather to lead them, as if the fine, powdery material that makes up the bands were emerging from the rump ends of the little worlds themselves. Additionally, color enhancement of the images suggested that all of the rings and all four moons are made up of the same ruddy, rusty-looking material. This led the scientists to postulate a whole new theory of ring formation.

Incoming micrometeorites, they figured, probably bombard the four innermost moons almost constantly. When the rubble strikes the satellites, it blasts large clouds of dust from their surfaces. Unimpeded by atmosphere, which the moons don't have, or gravity, which they do have but only barely, the dust simply drifts into space where it follows behind the satellites in a ghostly wake. Though the rings are diffuse enough to disperse relatively quickly, no sooner do they start to dissipate than they are replenished by other micrometeorite hits, releasing other clouds of dust. As long as the tiny bits of cosmic gravel keep colliding with the satellites—which ought to be as long as the Jovian system exists—the rings will continue to circle the planet.

Orbiting just above the small, ring-producing moons, of course, are the large Jovian moons: Io, Europa, Ganymede, and Callisto— the satellites discovered by the eponymous Galileo himself. While the rings and the small moons earned some of the scientists' attention, it was those large moons they wanted their new Jupiter probe to study most closely. Once again, it was those large moons that wound up paying the true scientific freight of the mission.

Callisto, the outermost of the quartet of large satellites, appeared to the spacecraft to have a fine and faint atmosphere, made principally of hydrogen, oxygen, and carbon dioxide. Though such a mixture of gases sounds altogether terrestrial, Callisto's atmosphere is nothing like Earth's, partly because it lacks nitrogen—which makes up 79 percent of the Earthly air blanket—and partly because it is infinitely more tenuous. What's more, while Earth's atmosphere was created by a combination of gases—some left over from the planet's original formation, others imported by comets, still others discharged by volcanoes—Callisto's impossibly wispy air is produced principally by molecules of water ice that are regularly knocked loose from the surface by solar radiation and then scattered into the Callistan sky where they swirl and recombine.

Though the moon has only the faintest atmosphere, it did appear to the spacecraft that Callisto might—like its sister Europa—have another, far more dramatic feature: a hidden ocean. Images and data from the *Voyager* probes suggested that of the four Galilean moons, it was Callisto that was far and away the least dynamic—its innards inert, its surface frozen, its crust carrying the scars of meteorite hits billions of years old. During *Galileo*'s encounters with the 2,976-mile Callisto, however, the ship detected what seemed to be a distinct magnetic field surging around the world, one never noticed by the more primitive *Voyager* instruments. While magnetic fields on other bodies are often created by a mobile, metallic core, density studies of the Callistan interior pretty much ruled such a structure out. Instead, the JPL team came up with a far more imaginative explanation for the moon's curious magnetism.

When Jupiter's own magnetic field collides with Callisto, it ought to pass right through the moon, washing over rocks, ice, or any other materials that lie beneath its surface. Most of that subsurface matter would not interact with the field in any significant way, but some of it—most notably salty water—would. If Callisto had a briny ocean, the magnetic field would set up an electrical current in it, much like the chemically driven charge that crackles inside an automobile bat-. tery. When this current surges through a body the size of Callisto, it would set up an equally vast magnetic field that would surround the moon and reach miles into space, where it could be easily detected by a passing spacecraft.

Why, however, should Callisto have an ocean? Without the tidal pumping that close-up moons like Io and Europa undergo, this most remote of the Galilean satellites simply would not have enough subsurface motion to melt its ice. The only way Callisto *could* warm itself up would be if it, like Jupiter, was rich in heat-producing radioactive materials. These, combined with the moon's natural gravitational pressure, might be enough to stoke Callisto's internal oven and liquefy at least part of its mantle.

Such a thermal process, if it were taking place at all, might not be confined to Callisto. *Galileo* images of nearby Ganymede confirmed earlier *Voyager* findings that this largest of the Jovian satellites has a surprisingly young crust, one that appears to have been partially melted and remelted repeatedly over its history. Just what was behind this process had never been clear, but if cold Callisto has a radioactive core and a liquid ocean, there was every reason to believe that Ganymede—with its greater bulk, greater gravitational energy, and periodically refreshed face—would, too. Long thought of as ice worlds, Callisto and Ganymede had become, overnight, possible water worlds, mere moons that all at once seemed to have all the chemical promise of planets.

Dynamic as these outermost of the four large Jovian satellites suddenly seemed, it was the two inner moons—Io and Europa—that once again proved to be the truly seething satellites in Jupiter's system. Though the vagaries of *Galileo*'s orbit have allowed it to make few close passes by Io, the spacecraft's cameras and other instruments have nonetheless been able to conduct at least cursory surveys of the moon's surface and have detected dozens of new volcanic vents smoldering there, some of which approach a scalding 1,000 degrees Fahrenheit. A number of eruption sites spotted in 1979 were still simmering in 1997, and showed every sign of continuing to percolate for years—perhaps millennia—more. During one five-month stretch between April and September 1997, a 250-mile-wide volcanic field appeared around the well-photographed Pillan Patera peak, blackening an area larger than the state of New Mexico.

Europa's face also proved to be more active than it seemed when *Voyager* flew by a generation earlier. Skimming just 125 miles over the moon in March 1997, *Galileo* was able to spot objects on its surface as small as twenty feet across. With the help of such sharp, long-distance vision, the spacecraft photographed relatively small iceberg-like objects all over the moon's frozen crust, fractured and capsized like ordinary terrestrial icebergs dotting ordinary terrestrial oceans. Unlike Earthly bergs, however, which bob and float in liquid water, Europan bergs did not appear to be in motion. Rather they seemed to be locked in place, almost as if they had suddenly become trapped when the water around them cooled and refroze. Superficially, this would seem to argue against an active liquid ocean, but on a surface as apparently turbulent as this one, the ice almost certainly thaws and freezes in cycles, perhaps never getting a chance to grow more than 3,000 feet thick before warming and liquefying again.

Such water as there is on Europa is apparently just as chemically rich as the JPL scientists always assumed it was. All over the moon's surface, *Galileo* spotted vast stains that bore the chemical signature of salt—particularly natron and Epsom salts, which form only in the presence of liquid water. As tidal forces squeeze Europa, briny water apparently squirts out of the moon like juice from an overripe tomato. Though the water evaporates quickly in the vacuum of space, the salt settles back down to the surface and remains there. If the smudges the eruptions leave behind are indeed as salty as they seem, there is even greater reason to believe that the moon is home not merely to an ocean but to a *living* ocean. It is not in clean, pristine, inland waters that the majority of Earth's aquatic life thrives, after all, but in the great salty soup of the planet's vast oceans. If such a rich medium breeds organisms on Earth, there is no reason it couldn't do the same on Europa, too.

Galileo was given a bit of extra time to explore that question. The spacecraft's principal mission was planned to run only from December 1995 to December 1997. When that two-year tour was finished, however, mission managers decided to extend it, sending the still-chugging ship on eight more flybys of Europa, four more of Callisto, and up to two more of smoking Io. It is only after these encounters that the Jet Propulsion Laboratory scientists will finally be finished with the hobbled old spacecraft, turning their attention at last to other ships and other missions.

What will happen to *Galileo* then is not entirely clear. It's possible that without Earthly navigators continuously monitoring the ship, the swirling gravity of Jupiter and its moons will combine to fling it out of the Jovian system altogether. In the alternative, it's possible that the pinball complexity of *Galileo*'s orbit will cause it simply to collide with one of the moons, annihilating the ship like an ordinary meteor. It's even possible that JPL will not merely abandon the spacecraft, but instead will eventually cede command of it to a university lab where apprentice engineers will gun its engines and change its headings, fooling around with the physics of orbital flight until the systems of the spacecraft wink out completely. Whatever becomes of *Galileo,* its truly useful, truly scientific work will at last be done, leaving the JPL researchers a legacy of data to study long after the ship itself expires.

When *Galileo* does die, it will almost certainly have left at least one great Jovian question—the question of life—unanswered. Settling that issue will take other spacecraft making other voyages out to the Jupiter system—specifically to Europa—and JPL appears to be getting ready to make those trips. Already in development in the Pasadena labs is a Europa orbiter spacecraft that may be ready to leave Earth and head out to the Jovian system as early as November 2003.

Engineers hoping to study Europa up close have always found the idea of placing a spacecraft into orbit around the moon a daunting one—even more daunting than the *Galileo* engineers found the prospect of trying to put a spacecraft into orbit around Jupiter. Jupiter's gravity, after all, is far greater than little Europa's, and the planet is thus able to hold onto a ship that is still moving at a pretty fast clip. In order to orbit Europa, an interplanetary spacecraft has to slow down to a comparative crawl of just a few thousand miles per hour, and that requires it to carry a virtual bazooka of an engine.

The Europa orbiter now in development is little *but* engine. Weighing no more than three or four adult human beings, the compact spacecraft is about three-quarters combustion chamber, exhaust bell, fuel tank, and fuel. It is only the remaining 25 percent of the ship's weight that is set aside for the sensors, cameras, and other hardware it is being built to transport in the first place. Modest as the mass of those instruments will be, the work they do will be considerable.

When the Europa ship enters orbit around the moon, it will be flying at an altitude of just 124 miles—significantly lower than the space shuttle usually orbits Earth and a bit closer than *Galileo* has ever come to Europa on any of its flybys. With upgraded cameras orbiting at that close a proximity, the ship should get the best images yet of the moon, photographing surface features significantly smaller than the jagged icebergs *Galileo* spotted. If any of those icebergs happens to be floating, if the ice around them happens to be cracking, if subsurface water happens to be welling up through those cracks, the spacecraft ought to notice.

Even if nothing so aquatically obvious is happening on the moon, the Europa orbiter should still be able to determine if an ocean lies beneath the surface ice. Along with its battery of cameras, the ship will also be carrying a planetary radar, an instrument that's never flown out to the Jupiter neighborhood before. Soaring through its barnstorming orbit, the spacecraft will beam radar signals down to the surface, looking for the telltale echo of liquid water sloshing beneath the ice. If the ship indeed picks up such signature signals, it will map the spots where the underground water appears to be located, flagging the photographs of those areas for especially close study by the engineers on the ground.

Also assisting in the spacecraft's ocean hunt will be a more sophisticated device known as a laser altimeter. As Europa moves through its orbits around Jupiter, the tidal flexing caused by the planet and the surrounding moons subtly distorts the shape of the world. Just how much it distorts it depends on what the moon itself is made of. If Europa is solid ice all the way through, it should flex by no more than three to six feet during any one orbit. If, however, there's an ocean beneath the ice, things will be a good deal more malleable, with the moon changing shape by as much as twenty to thirty feet. Beaming a needle-thin laser signal down to the surface, the Europa orbiter should be able to measure the moon's shape to within just a few inches. Keeping track of the Europan flexing as the spacecraft glides through its orbits—and correcting for surface irregularities that could throw the readings off—the JPL scientists should be able to take some all but flawless measurement of the moon and draw some all but flawless inferences about its innards.

If those innards do look liquid, JPL has bigger plans still for Europa. Already being developed by the Pasadena engineers is another spacecraft that would not simply circle the Jovian moon, but land on it. Flying out to Europa and setting itself down on a patch of ice that the radar and laser suggested was especially thin, the ship would release a four-foot cylindrical probe with a heated tip that would melt through the frozen crust looking for water. Should it find any—even a body as small as an underground lake or pond—it would release a smaller, one-foot, submarine-like probe into the water, which would propel itself around, sampling the marine chemistry and looking for signs of life.

Yet another, equally improbable spacecraft might be sent on a mission to bring a tiny bit of Europa home. Launched on a simple flyby trajectory, the sample-return ship would approach Europa and fire a twenty-pound, shot put–like sphere out into space. The cosmic cannonball would arc down toward the surface of Europa, strike the frozen crust with the force of a dynamite blast, and release a mushroom cloud of ice particles into space. The slower-moving mother ship would then fly through this crystalline mist, collect a bit of it, and carry it back to Earth for analysis in the lab.

Whether any of these future missions will actually fly, of course, is by no means certain. If the tortured history of projects like *Voyager* proves anything, it's that a lot can happen between the time a spacecraft is proposed and the time it's built, and no mission should be considered a sure thing until the ship has actually left the pad. Even as the Europa probes are still being debated and designed, however, another JPL spacecraft—the *Cassini-Huygens* probe—is already speeding silently through the solar system, making its way toward another, even more distant destination: Saturn and its atmosphere-shrouded moon Titan.

Conceived shortly after the *Voyager* flybys of Saturn in 1980 and 1981 but not launched until October 1997, *Cassini* is modeled after its sister ship, *Galileo*—with a few decided improvements. There is no tangle-prone, snap-prone tape recorder on this spacecraft; any memories the ship needs to preserve are stored in a digital system that records data largely without moving parts. Similarly, there is no collapsible—and fallible—rib-and-mesh antenna. When *Cassini* left the ground, it left not inside the cramped cargo bay of a space shuttle, but inside the roomy payload area of a Titan-Centaur rocket. The more spacious missile meant more room for an antenna, and *Cassini* carried a dilly: a 13.1-foot hard-shell dish that was molded in a single unbreakable piece, in a single open position.

The similarities between *Cassini* and *Galileo*, however, outnumber their differences. Like its Jupiter-touring predecessor, the *Cassini-Huygens* ship is actually two ships: an orbiter and a probe. The main body of the spacecraft—named after Giovanni Cassini, the seventeenth-century astronomer who discovered four Saturnian moons and the gap in the planet's rings—was designed to orbit Saturn and spend at least four years photographing and studying it. Making no fewer than sixty laps around the world, the 2.5-ton *Cassini* will swoop to within 110,000 miles of Saturn's cloudtops, training its cameras and other instruments on the planet itself, its array of rings, and at least eight of its well-studied seventeen moons: Mimas, Enceladus, Tethys, Dione, Rhea, Titan, Hyperion, and Iapetus. The ship may also get at least a glimpse of a little-known eighteenth moon—tiny, twelve-mile-wide Pan, spotted on vintage *Voyager* photos in 1990.

Several months after *Cassini* goes into orbit around Saturn in January 2004, the *Huygens* probe attached to the exterior of the ship—named after Christiaan Huygens, who discovered Titan in 1655—will set to work. Measuring 8.9 feet across and weighing 770 pounds, *Huygens* is a blunt, bullet-shaped assembly modeled after the atmospheric instrument package carried aboard *Galileo*. Unlike the *Galileo* probe, however, this one was not built to study the chemistry of a planet; instead, it will take its plunge through the rust orange air of Titan.

Released from the speeding *Cassini* orbiter, the *Huygens* probe will navigate its way to the nearby moon and slam into its dense air blanket at a bone-jolting 13,750 miles per hour. Free-falling through the surrounding haze, the probe will encounter greater and greater atmospheric resistance until, after just three minutes, its velocity will have slowed to a comparatively sluggish 895 miles per hour. At that point, a 6.5-foot drogue parachute will pop from the pod, followed shortly after by a large, billowing twenty-seven-foot main chute. The *Huygens* probe will continue to descend through the opaque atmosphere for more than two and a half hours, its speed steadily slowing, the already dim light around it growing murkier still, until finally, when it is drifting downward at just 15 miles per hour, it will bump down on the Titanian surface. Just what the probe will find there depends on just where it lands.

Though no camera that processes images in visible light has ever been able to penetrate Titan's heavy atmosphere, cameras that work in the infrared can peer right through it. The Hubble space telescope, which has been orbiting Earth since 1990, is equipped with just such a camera, and not long ago it trained its gaze on Titan and spotted what appeared to be a curiously bright patch covering a portion of the moon's surface. If a formation that big were discovered on a world like Earth, scientists would guess it was an ocean or at least a large lake, and on the surface of Titan, it appears that that might be exactly what it is.

In the paralyzing cold that dominates Titan's climate, an ocean made of liquid water clearly could not exist. What some scientists think the Hubble telescope spotted instead is a rich mixture of liquid ethane and methane. Two of the signature chemicals of organic science, ethane and methane are the surest signs that Titan may indeed be a prebiological place. Should the *Huygens* probe land in this vast pool, accelerometers and float detectors attached to the buoyant machine will measure how much of it stays above the surface of the liquid and how much of it becomes submerged as it floats this way and that. Factoring this information together with the known mass of the probe should help scientists determine the density of the liquid, and that, in turn, should help them determine its precise composition.

Even before the buoyancy sensors can get to work, however, sci-entists should have a pretty good idea of whether *Huygens* touches down on land or in liquid. Built into the body of the probe is an or-dinary microphone that will provide mission controllers with the de-cidedly *extra*ordinary ability to listen in on their ship. As the *Huygens* pod descends, JPL engineers will be able to hear the atmo-sphere screaming by, the chutes popping open, the winds pushing the little instrument package through the orange sky. If the probe indeed does land in the ethane-methane ocean, they should also hear the thump, the slosh, and then—implausibly—the soft slapping of the Titanian waves as they lap against the sides of the interplanetary buoy.

Wherever *Huygens* lands, it will have to do its work fast. With solar panels unworkable in the Titanian gloaming, radioactive gener-ators producing too much heat for so small a craft, and fuel cells too unreliable, the only power source the probe can carry is a battery— one that will have a life span of just thirty to sixty minutes once the instrument package is on the surface. Throughout that brief period, the probe's numerous sensors will pull on that battery hard. Visual and infrared cameras will scan the landscape, peering out, down, and even up, in an attempt to determine how much feeble sunlight makes it through the blanket of clouds; a spectral radiometer and mass spectrometer will gather in sips of air and analyze its chemistry; a Doppler wind instrument will study the local breezes; an aerosol collector will examine liquid and solid particles suspended in the at-mosphere. All of the findings from all of these sensors will be trans-mitted up to the *Cassini* mother ship orbiting overhead and then relayed the eight-tenths of a billion miles back to Earth. Less than an hour after the first data bit is sent out from the probe, the instru-ments as a group will flicker and die and the *Huygens* lander will fall forever silent. Unlike the *Galileo* probe, however, which was vapor-ized by the world it came to visit, the *Huygens* probe will endure for centuries, a metal mile-marker commemorating the farthest spot hu-manity has reached in its decades-long effort to place its machines on worlds not its own.

Once Titan is studied up close, JPL's preliminary look at the large bodies of the known solar system will at last be complete—or almost complete. While seven of the other eight planets and all of their large moons have at least been approached by Earthly machines, one planetary system—Pluto's—has gone utterly unexplored. It was in 1930 that Clyde Tombaugh, an astronomer at Lowell Observatory in Arizona, first spotted an impossibly faint dot lost among the far brighter stars of the wintertime sky. Tracking the dot for more than a year, he found to his surprise that it was moving—far more slowly than a speedy planet like, say, Mercury or Venus, but far faster than the stars themselves, which really oughtn't move at all. The dot, it turned out, was a ninth planet, a rocky, icy, 1,423-mile-wide world orbiting the sun nearly 3.7 billion miles from Earth. Nearly half a century later, in 1978, James Christy, an astronomer working at the United States Naval Observatory, discovered that tiny Pluto is itself circled by a tiny moon, a 737-mile ball, also made up principally of rock and water ice. The planet, sunk in the permanent darkness of deep space, was named Pluto, after the overseer of the underworld. Its moon was named Charon, the boatman said to ferry condemned souls across the River Styx.

By any measure, Pluto and Charon are an unusual pair. First of all, Charon is—in relative terms, at least—huge, measuring more than half the size of its parent world. Such a one-to-two moon-to-planet ratio is far and away the biggest in the solar system. Stranger still is the way the two bodies move. While Earth's moon orbits its parent planet at a mean distance of 238,000 miles and other large moons like Ganymede, Callisto, and Titan orbit at three or four times that altitude, Pluto and Charon spin in a close-up embrace just 12,000 miles apart, or only a bit more than the air-mile distance between Tokyo and Rio de Janeiro. To the untrained eye, Charon would not appear to be Pluto's satellite at all; rather, the two arm's-length worlds would seem to be orbiting each other, twirling about a common axis like the spherical weights at the opposite ends of a set of spinning barbells.

If the path Charon inscribes about Pluto is an uncommon one, the path both bodies inscribe around the sun is even more so. For the most part, the local planets maintain reasonably orderly solar orbits, moving about their anchor star in roughly circular paths that remain roughly in the plane of the solar system. Pluto's and Charon's orbit, by contrast, is an asymmetrical mess. Like the irregularly orbiting captured moons that circle the larger planets, the ninth planet and its companion satellite move about the sun at a dizzyingly inclined angle, rising 17.15 degrees above the solar equator and plunging 17.15 degrees below it. Additionally, the orbit is not even remotely circular. Though Pluto maintains an average distance of nearly 3.7 billion miles from the center of the solar system, at times it closes to within 2.8 billion miles, actually bringing it inside Neptune's orbit for twenty out of the 248 years it takes to travel around the sun.

For this reason more than any other, astronomers who have always thought of Pluto as a fully credentialed planet have lately begun to conclude that it might be nothing of the kind. Rather, like Neptune's Triton, Saturn's Phoebe, and all eight of Jupiter's outer satellites, Pluto and Charon might be nothing more than two bits of free-floating debris that went wandering through the solar system and became trapped in orbit around one of its bodies. If the planet and its moon are indeed nothing more than a pair of cosmic strays, it is almost certain that they came from the same rich river of detritus that gave birth to Triton, Nereid, and most of the solar system's comets: the Kuiper Belt. Superficially, Pluto and Charon certainly seem to be made of Kuiper Belt stuff—a bit of rock, lots of ice, a nimbus of haze surrounding them both. What's more, they move with Kuiper Belt motion. If their trajectories were to change by only a few degrees, the two bodies would not be orbiting the sun like planets at all, but swooping around it like comets, plunging toward the center of the solar system, flaring brightly for a few years, and then speeding back out again.

Formulating theories about the origin and composition of Pluto and Charon, of course, is not the same as proving those theories, and for years, astronomers have known that the only way to determine anything for sure is to fly out to the edge of the solar system and look up close. More than twenty years ago the outer planets offered to help out, moving into a tidy line that would have easily allowed a

Voyager spacecraft launched to Jupiter to make a sharp gravitational turn and speed out to Pluto. But the stewards of the ship decided not to make that trip, and the planets, indifferent to that choice, glided back out of position. Visiting Pluto now would require a whole new journey flown by a whole new kind of spacecraft. Even now, that spacecraft and that mission are being planned.

Sometime in the early winter of 2004, the way JPL envisions things, the lab will launch a small featherweight spacecraft dubbed the *Pluto-Kuiper Express.* Weighing as little as 500 pounds and carrying a dish antenna measuring no more than six feet across, *Pluto-Kuiper* will easily be the smallest deep-space machine ever built. Such a light ship can move at a considerable clip, and less than two years after its launch, it should have flown out as far as the neighborhood of Jupiter. Arcing partway around the planet, it will pick up a gravitational boost and fling itself in the direction of Pluto, arriving there sometime in early 2013 and flying directly through the 12,000-mile gap that separates the planet and its moon.

Assuming mission planners are able to pull off this feat of navigational needle-threading, the spacecraft should reward their efforts amply. Orbiting where Pluto does, the planet is 39 times farther from the sun than Earth is. Given the squared geometry of solar illumination, that means it receives 961 times less solar light, with the sun appearing to be little more than a very bright star, all but lost against a background curtain of other stars. Such a tiny stellar spangle does not provide much illumination, and taking pictures of a planet so feebly lit would not be easy. The *Voyager* probes overcame a similar problem in a relatively crude way—by leaving their cameras' shutters open, collecting as much light as they could, and pivoting the main body of the spacecraft in any way necessary to keep the picture from smearing.

Pluto-Kuiper Express will require no such complicated maneuvering. Built into the body of the ship will be a system that reads light not a beam or stream at a time, but a single photon at a time. Pouring into the imaging system, these massless subatomic units—essentially particles of light—strike electrons on a semiconductor, release a charge that excites a picture element, and then build up an image one fraction of a pixel after another. Systems like this—known as charge-coupled devices, or CCDs—work particularly well in extremely low light environments in which the detectors run little risk of being swamped by too much incoming illumination. They are also especially well adapted to cold, since heat can overexcite the hardware as easily as light, knocking loose electrons and creating false light and shadow where none exists. The deep black, bitterly cold Pluto environment, where temperatures approach −400 degrees, was practically made for a CCD.

When the *Pluto-Kuiper Express* arrives at its destination, its own CCD will not have much time to operate. The spacecraft will be able to photograph the tiny planet and moon for only about 3.2 days as it approaches them and another 3.2 days as it sails away. As it happens, however, this 6.4-day period is just a bit longer than a Pluto day, meaning that while the ship is in the vicinity of the Plutonian system, it will be able to watch most of the planet rotating beneath it, capturing images of both of its hemispheres as they turn slowly through space. Any feature on the face of the world that's at least as big as, say, a sports stadium will be faithfully recorded by the cameras.

Whatever pictures the *Pluto-Kuiper* ship does collect should be the last anyone sees of the surface of Pluto for a long time. For the last twenty years the planet has conveniently been in the portion of its orbit that brings it closest to the sun, having crossed inside the orbit of Neptune in January 1979 and crossed back out in February 1999. That relatively small change in proximity to the warmth of the solar fires is enough to heat up the Plutonian atmosphere slightly, keeping it gaseous and transparent. As Pluto moves farther away, however, once again becoming the ninth planet for the remaining 228 years of its orbit, the atmosphere will chill, freeze, and settle onto the surface in an opaque shell. When the *Pluto-Kuiper Express* reaches the planet, JPL scientists believe the air will still be vaporous enough for the spacecraft to see through. Any ships that are launched toward Pluto in the next two centuries, however, will have a far harder time of things, encountering only a flash-frozen globe, completely hidden by solidified gases from even the sharpest prying eyes.

Just where NASA is headed after it reconnoiters Pluto, after it returns to Europa, after it sails about Saturn and fires a probe into Titan, is unsettled for now. Those projects alone, of course, could keep the agency busy for the better part of a generation. Whatever ships will follow along after is for later generations to determine. But deep-space probes are not the only way to explore the cosmos, and even without sending a machine so much as a mile off Earth, scientists are discovering that the solar system still has new sides of itself to reveal.

Late in the summer of 1997, astronomers working in the mountainous Palomar Observatory in southern California decided to spend an evening or two studying the planet Uranus. Their telescope was a good one—a fat, fifteen-foot-wide reflector equipped with its own sensitive CCD. Such a sharp-eyed system was unlikely to miss any of the light Uranus reflected back to it, and indeed, the images the telescope collected were crisp ones. The astronomers could see the sea blue sphere of the planet itself—unchanged from the time eleven years earlier when *Voyager 2* glided serenely by. They could see the spangle of fifteen moons spinning like electrons around the world. They could watch as the moons orbited along in their neat ranks, all of them remaining in the equatorial plane of Uranus, all of them moving a uniform counterclockwise.

That, however, wasn't all the images revealed. Hovering far above Oberon, the highest flying of the Uranian moons, were two other shimmery points. Both of them appeared to be vaguely red in color; neither of them appeared to be even remotely round in shape. One of the two points—the lower-flying one—looked to be a good 4.5 million miles above the surface of Uranus; the other was a good 7.5 million. The astronomers took other pictures of the points and noticed that both of them appeared to be moving in an arc-like trajectory; more images still and the arcs became elongated circles, ones that took them in a high, elliptical orbit completely around the planet. That final finding, the scientists knew, changed things entirely. The two inconsequential points weren't just points at all, but two previously undiscovered moons, the first spotted anywhere in the solar system since the last image from the *Voyager*s came flickering back to Earth years earlier.

The new Uranian satellites—measuring 37 and 75 miles across— were not much to speak of: organics-rich cosmic BBs that probably flew in from the deeper cosmos and got caught in the planet's gravitational net. But while they might have been mere flying rocks once, they were undeniably moons now, and as such, they earned the right to be named. The smaller of the two satellites was accordingly dubbed Caliban, the sorrowful slave of the magician Prospero in Shakespeare's *The Tempest*. The larger one was named Sycorax, Caliban's sorceress mother.

With their discovery, Caliban and Sycorax brought the solar system's moon total to a robust sixty-three—or sixty-two more than the astronomical community ever imagined before Galileo Galilei began looking skyward nearly four centuries ago. If scientific history indicates anything at all, it is that even so great a satellite census could grow further still. Just who will count the new worlds, however, is impossible to say. It was the long-ago ships of the Jet Propulsion Laboratory that discovered so many of the sixty-three known moons. It was the long-ago scientists of earlier eras who discovered most of the rest. Now, it will be up to new ships—and to new observers peering up from the ground—to find any more.

Epilogue

The people in the room were trying hard not to spill food on all the spacecraft standing around them. It wasn't easy, what with hot hors d'oeuvres and cold hors d'oeuvres and little iced drinks to juggle. Already at least one person in the crowd had thumped into the side of a V-2 rocket, nearly sloshing a bit of soda down its metal skin. But while a V-2 could probably take the abuse, most of the rest of the machines—particularly the elegant *Ranger* moonship—were far more fragile pieces of hardware.

Or at least the people here today thought of them as fragile. But then, they *would* think of them that way. Three decades ago, deep in the early 1960s, the aristocratic-looking gentlemen in the room had had to build six of the spindly *Rangers*, only to see them fall apart in their hands, before a seventh one finally operated the way it had been designed to operate. Jim Burke was among those engineers here today; so were Bud Schurmeier and Cliff Cummings, Burke's former bosses. Also here, arriving just a few minutes ago, was William Pickering, smiling, glad-handing, methodically working the room with a calm and clarity that belied his eighty-plus years.

The place all of these people had come today was the Smithsonian Institute's Air and Space Museum in Washington, D.C. And the reason they had come was to pay tribute to Jim Webb, the long-ago administrator of NASA. It had been more than two decades since Webb had left the agency, and in all that time nobody had thought to commemorate his tenure. In 1968 he had been allowed simply to empty out his desk, gather up his belongings, and slip quietly back into private life. Now, at last, someone in the protocol office had decided to rectify that oversight.

Hundreds of people from most of the NASA facilities had been invited here for the event today, including the men and women of the Jet Propulsion Laboratory in Pasadena. Pickering, who himself had left the agency years before, had been among the earliest of the former JPL scientists to answer the invitation, and to no one's surprise, he had answered yes. With word that the lab's first chief was going to attend, most of the other people who had worked for him decided to show up, too.

If JPL's scientists were well represented at the Air and Space Museum today, JPL's machines were even more so. Displayed around the place like so many animal heads from so many wilderness expeditions were the *Rangers*, *Surveyors*, *Vikings*, *Mariners*, *Voyagers*, and other ships that had, over the years, traveled to the moon, Mars, Jupiter, Saturn, Uranus, Neptune, and elsewhere. Standing nearby, dwarfing both these spacecraft and the men who built them, were the boosters that had started all the interplanetary ships on their journeys. Rising up toward the glass ceiling of the museum's atrium was an arsenal of Agenas, Atlases, Junos, Redstones, and Vanguards, all long since drained of their fuel and stripped of their vitals—all, essentially, taxidermized—built for flight but destined never to fly anywhere at all.

The event this evening played out over a few hours, and throughout that time Pickering had never had the opportunity to do more than briskly shake the hand of Webb, the man he was ostensibly here to honor. A long time ago, of course, he did not feel quite as warmly toward his onetime boss—not when all the *Rangers* were failing and the press was complaining and Webb was threatening to have Pickering's head if the faltering moonships didn't start flying right. But those years were past, and tonight Pickering would not mind finding a few minutes to reflect with Webb on happier times. Finally, toward the end of the night, as most of the other guests were gathering their coats and moving toward the exits, the former JPL chief noticed the former NASA chief standing quietly by himself near one of the old, hollow boosters. Pickering walked over and joined him.

"It was a pleasant evening," he said simply.

Webb nodded.

"Long overdue," Pickering added.

Webb shrugged, but with a small smile.

Pickering looked around the room slowly. Above his head, hung from cables attached to ceiling beams, were three of his spacecraft, suspended in eternal flight. Webb followed his gaze.

"Nice machines," the ex–NASA administrator said after a silence.

Now it was Pickering's turn to shrug. "They did the job," he said.

Pickering and Webb fell quiet again, and after a moment the muffled click of a circuit breaker sounded at the other end of the gallery and a bank of ceiling lights switched off, plunging part of the display into darkness. Another bank of lights followed, then another. The few people left in the museum started moving toward the exits. Wordlessly, the two scientists began to follow them. Before they had walked even a few steps, however, Webb stopped and turned to Pickering.

"You know," he said, "I may not have mentioned it at the time, but I always thought you fellows had a fine operation out there."

Pickering nodded.

"A real fine operation," Webb repeated. Pickering smiled and the two men walked on.

Overhead, a few more lights clicked softly off.

Appendix 1

The Sixty-three Moons

Name	Altitude (Miles)	Diameter (Miles)	Orbit (Days)	Discovery (Year)
Earth				
The Moon	238,000	2,155	27.3	—
Mars				
Phobos	5,815	13.6	.32	1877
Deimos	14,545	8.7	1.3	1877
Jupiter				
Metis	79,360	25	.20	1979
Adrastea	79,967	12	.29	1979
Amalthea	112,406	117	.49	1892
Thebe	137,578	62	.67	1979
Io	261,392	2,251	1.8	1610
Europa	415,958	1,945	3.5	1610
Ganymede	663,400	3,262	7.1	1610
Callisto	1,167,469	2,976	16.7	1610
Leda	6,878,280	10	239	1974
Himalia	7,117,600	115	251	1904
Lysithea	7,266,400	22	259	1938
Elara	7,276,940	47	260	1905
Ananke	13,144,000	19	631	1951
Carme	14,012,000	25	692	1938
Pasiphae	14,570,000	31	735	1908
Sinope	14,694,000	22	758	1914

Name	Altitude (Miles)	Diameter (Miles)	Orbit (Days)	Discovery (Year)
Saturn				
Pan	82,820	12	.57	1990
Atlas	85,355	19	.60	1980
Prometheus	86,397	62	.61	1980
Pandora	87,854	56	.62	1980
Epimetheus	93,880	74	.69	1966
Janus	93,911	118	.69	1966
Mimas	115,022	243	.94	1789
Enceladus	147,572	310	1.4	1789
Tethys	182,689	657	1.89	1684
Telesto	182,689	19	1.89	1980
Calypso	182,689	16	1.89	1980
Dione	233,988	694	2.73	1684
Helene	233,988	20	2.73	1980
Rhea	326,764	949	4.5	1672
Titan	757,534	3,193	15.9	1655
Hyperion	918,282	180	21.2	1848
Iapetus	2,208,006	905	79.3	1671
Phoebe	8,030,240	136	550	1898
Uranus				
Cordelia	30,846	16	.34	1986
Ophelia	31,473	20	.38	1986
Bianca	36,682	27	.43	1986
Cressida	38,295	41	.46	1986
Desdemona	38,848	36	.47	1986
Juliet	39,901	52	.49	1986
Portia	40,980	68	.51	1986
Rosalind	43,354	33	.56	1986
Belinda	46,658	42	.62	1986
Puck	53,323	95	.76	1985
Miranda	80,505	292	1.4	1948
Ariel	118,375	718	2.5	1851
Umbriel	164,906	725	4.1	1851

Name	Altitude (Miles)	Diameter (Miles)	Orbit (Days)	Discovery (Year)
Titania	270,489	978	8.7	1787
Oberon	361,721	943	13.4	1787
Caliban	4,464,000	37	930	1997
Sycorax	7,564,000	75	1,280	1997
Neptune				
Naiad	29,902	36	.29	1989
Thalassa	31,043	50	.31	1989
Despina	32,586	92	.33	1989
Galatea	38,409	98	.42	1989
Larissa	45,601	119	.55	1989
Proteus	72,936	257	1.1	1989
Triton	219,976	1,674	5.9	1846
Nereid	3,418,060	210	360	1949
Pluto				
Charon	12,176	737	6.3	1978

Appendix 2

The Unmanned Missions to the Moons

Spacecraft	Date Launched	Destination	Results
Ranger 1	August 23, 1961	The moon	Failed to leave Earth orbit
Ranger 2	November 18, 1961	The moon	Failed to leave Earth orbit
Ranger 3	January 26, 1962	The moon	Missed moon by 23,000 miles
Ranger 4	April 23, 1962	The moon	Hit far side of moon; returned no pictures
Ranger 5	October 18, 1962	The moon	Missed moon due to on-board power failure
Ranger 6	January 30, 1964	The moon	Successfully crashed on lunar surface; camera blackout prevented return of pictures
Ranger 7	July 28, 1964	The moon	Successfully crashed in the Sea of Clouds; returned 4,316 pictures

Spacecraft	Date Launched	Destination	Results
Ranger 8	February 17, 1965	The moon	Successfully crashed in Sea of Tranquillity; returned 7,137 pictures
Ranger 9	March 21, 1965	The moon	Successfully crashed in crater Alphonsus; returned 5,814 pictures
Surveyor 1	May 30, 1966	The moon	Soft-landed in the Ocean of Storms; returned over 10,000 pictures
Lunar Orbiter 1	August 10, 1966	The moon	Entered lunar orbit; returned 207 pictures
Surveyor 2	September 20, 1966	The moon	Engine failure caused crash on lunar surface
Lunar Orbiter 2	November 6, 1966	The moon	Entered lunar orbit; returned 211 pictures
Lunar Orbiter 3	February 4, 1967	The moon	Entered lunar orbit; returned 211 pictures
Surveyor 3	April 17, 1967	The moon	Soft-landed in the Ocean of Storms; returned 6,315 pictures

Spacecraft	Date Launched	Destination	Results
Lunar Orbiter 4	May 4, 1967	The moon	Entered lunar orbit; flew over south pole; returned 193 pictures
Surveyor 4	July 14, 1967	The moon	Undetermined malfunction caused crash on lunar surface
Lunar Orbiter 5	August 1, 1967	The moon	Entered lunar orbit; returned 212 pictures
Surveyor 5	September 8, 1967	The moon	Soft-landed in the Sea of Tranquillity; returned over 18,000 pictures
Surveyor 6	November 7, 1967	The moon	Soft-landed in moon's Central Bay; returned over 29,500 pictures
Surveyor 7	January 7, 1968	The moon	Soft-landed in Tycho crater; returned 21,274 pictures
Mariner 9	May 30, 1971	Mars	Orbited Mars; flew by Phobos; returned 7,329 pictures
Pioneer 10	March 3, 1972	Jupiter	Successfully flew through Jovian system

Spacecraft	Date Launched	Destination	Results
Pioneer 11	April 5, 1973	Jupiter and Saturn	Successfully flew through Jovian and Saturnian systems
Voyager 2	August 20, 1977	Jupiter, Saturn, Uranus, and Neptune	Launched before *Voyager 1;* successfully flew through all four systems
Voyager 1	September 5, 1977	Jupiter and Saturn	Successfully flew through Jovian and Saturnian systems
Galileo	October 18, 1989	Jupiter	Currently orbiting planet
Cassini-Huygens	October 15, 1997	Saturn	Currently en route

Author's Note

Just as it takes a small army of willing people to send a spacecraft to its cosmic destination, so too does it take a smaller, more modest army of people to tell the tale of the mission that ship flies. I am deeply indebted to just such an army for making *Journey Beyond Selēnē* possible.

Most of the stories recounted here were reconstructed through interviews with the scientists, administrators, and engineers who have been flying the spacecraft of the Jet Propulsion Laboratory for the past forty years. The rest were researched with the help of thousands of pages of documents and news clips, all of which have been preserved in NASA's various libraries and all of which were graciously made available to me. Conversations that appear in quotes in the text were reconstructed through interviews with one—and often more than one—of the principals involved.

Among those scientists and engineers who gave most generously of their time to help me tell this tale, the one to whom I owe the greatest thanks is Dr. William Pickering—thinker, visionary, and New Zealand's gift to America. May we someday contrive a way to repay the debt. Also lending enormous help were the gifted and engaging Rich Terrile and Torrence Johnson of JPL; my thanks for all the times they took my calls and all the times they explained to me—patiently—how the solar system works. Appreciation goes as well to countless other folks at JPL and elsewhere in the space community, including Jim Burke, Karen Buxbaum, John Casani, Clifford Cummings, John Delaney, Eugene Giberson, Norm Haynes, Linda Morabito Hyder, Ray Heacock, Charles Kohlhase, Richard Laeser, Victoria Melikan Lairmore, Matt Landanow, Peter Lyman, Chris McKay, Dennis Madsen, Ellis Miner, Bill Muehlberger, Marcia Neugebauer, Bill O'Neil, Toby Owen, Robert Parks, Richard Rudd, Bud Schurmeier, Paul Spudis, Steven Squyres, Rob Staehle, David Stevenson, Steve Synnott, Ewen Whitaker, and Don Wilhelms. Special thanks are also owed to Dave Scott, a man uniquely qualified to describe any moon.

A number of fine authors and their fine books also assisted me in my research; among them: *Galileo's Commandment*, edited by Edmund Blair Boles; *Exploring Space*, by William E. Burrows; *A Man on the Moon*, by Andrew Chaikin; *Lunar Impact: A History of Project Ranger*, by R. Cargill Hall; *The Story of Astronomy*, by Lloyd Motz and Jefferson Hane Weaver; *Astronomy and Cosmology*, by John North; *The Once and Future Moon*, by Paul Spudis; *Voyager Tales*, by David W. Swift; *To a Rocky Moon*, by Don E. Wilhelms; and *Astronomy Through the Ages*, by Robert Wilson.

In addition to these people, numerous others in the publishing and journalism community lent their help. Thanks to Philip Elmer-DeWitt of *Time* magazine, for providing the kind of flexible environment in which it's possible to write both a news story and a book, and do a creditable job of both; thanks as well to Michael D. Lemonick, also of *Time*, for his counsel and commiseration. Much appreciation also goes to Alice Mayhew and Roger Labrie of Simon & Schuster, and to everyone at the Joy Harris Literary Agency, especially Leslie Daniels and the infinitely supportive, infinitely patient Joy Harris.

Finally, thanks and much love to Splash, Steve, Garry, and Bruce Kluger, Lori Oliwenstein, and Alene Hokenstad, the people with whom I have long shared my orbit; to Richard Kluger for setting an incandescent example; to Bridgette, Emily, Audrey, and Mateo, for reminding me that the universe can still be a place of enchantment; and finally, to my wife, Alejandra López Kluger, for getting up early to watch Venus rise with me—and for all the things that implies.

Index

About the Author

Jeffrey Kluger is a senior writer at *Time* magazine covering science and the space program. He is the coauthor, along with astronaut Jim Lovell, of *Lost Moon,* the book that served as the basis of the 1995 movie *Apollo 13.* He is also the author of *The Apollo Adventure,* which accompanied the release of the movie. His features and columns have appeared in *The New York Times Magazine, GQ,* and *Newsday,* among other publications. He lives in New York City.